Bonsai

A Complete Guide to Master the Ancient Art

(Discover How to Nurture Your Own Miniature Tree to Improve Your Well-being)

Billy Lorusso

Published By **Oliver Leish**

Billy Lorusso

All Rights Reserved

Bonsai: A Complete Guide to Master the Ancient Art (Discover How to Nurture Your Own Miniature Tree to Improve Your Well-being)

ISBN 978-1-7780617-6-9

No part of this guidebook shall be reproduced in any form without permission in writing from the publisher except in the case of brief quotations embodied in critical articles or reviews.

Legal & Disclaimer

The information contained in this book is not designed to replace or take the place of any form of medicine or professional medical advice. The information in this book has been provided for educational & entertainment purposes only.

The information contained in this book has been compiled from sources deemed reliable, and it is accurate to the best of the Author's knowledge; however, the Author cannot guarantee its accuracy and validity and cannot be held liable for any errors or omissions. Changes are periodically made to this book. You must consult your doctor or get professional medical advice before using any of the suggested remedies, techniques, or information in this book.

Upon using the information contained in this book, you agree to hold harmless the Author from and against any damages, costs, and expenses, including any legal fees potentially resulting from the application of any of the information provided by this guide. This disclaimer applies to any damages or injury caused by the use and application, whether directly or indirectly, of any advice or information presented, whether for breach of contract, tort, negligence, personal injury, criminal intent, or under any other cause of action.

You agree to accept all risks of using the information presented inside this book. You need to consult a professional medical practitioner in order to ensure you are both able and healthy enough to participate in this program.

Table Of Contents

Chapter 1: Bonsai Basics 1

Chapter 2: Ficus Bonsai Care And Maintenance.. 14

Chapter 3: Fukien Tea Care And Maintenance.. 24

Chapter 4: Cotoneaster Care And Maintenance.. 35

Chapter 5: Pines Care And Maintenance 45

Chapter 6: Buy Or Create From Scratch? 55

Chapter 7: The Basics Techniques 75

Chapter 8: Designing Your Bonsai 85

Chapter 9: Different Styles Of Bonsai..... 98

Chapter 10: Principal Resources........... 122

Chapter 11: The Bonsai Art 131

Chapter 12: Selecting The Best Bonsai . 138

Chapter 13: Potting And Repotting As Described.. 147

Chapter 14: Tuning In To The Nature's Rhythms Using Seasonal Care 155

Chapter 15: A Source Of Serenity 166

Chapter 1: Bonsai Basics

Bonsai and garden art that is rooted in long-standing tradition, provides an exciting glimpse into the harmonious relationship between human and nature. The word "Bonsai" originates from the Japanese word "bon" (meaning tray) as well as "sai" (meaning planting) which refers to the careful cultivation of miniature trees inside small containers. The living statues are a testament to the beauty of nature as well as humans' hand in caring for and sculpting the trees.

But, the idea of reducing trees into miniatures to make Bonsai does not have to be solely Japanese from the beginning. The roots of the concept are traced to the early days of China and similar techniques have been practiced since a century long. As time passed, the art was able to transcend borders and evolved distinctive styles and methods in various different styles and cultures.

Bonsai involves more than cultivating small trees. It is the profound philosophy that has its roots in the principles of the simplicity of balance, harmony and harmony. The practice invites the participants to be in dialogue with the natural world and cultivates a bond which goes beyond the physical practice of cultivating. Through Bonsai practitioners, people gain an appreciation for the passing of time because these tiny trees capture all the years in tiny shapes.

It is said that the art of Bonsai offers a fascinating trip, in which the endless humanity's creativity meets the elegance and strength of nature. Through this journey, we dive into the intricate web of Bonsai and explore its past as well as its philosophical tenets, as well as the skill required for the creation and nurturing of the living artifacts.

Brief History of Bonsai

The gorgeous artwork of Bonsai is awe-inspiring, thanks to its exquisite miniature trees and stunning aesthetics, is a part of a culture that spans centuries and crosses continents. To fully enjoy Bonsai you must go to explore its fascinating history.

Bonsai's roots are traced to the beginning of China in the time of growing small pots of trees, referred to by the name of "penzai," flourished over 1000 years long. The early Chinese people tended to various kinds of trees, and would often display the trees in indoor spaces as a symbol of elegance and respectability. The miniature trees weren't solely admired for their decorative splendor, but also because of their significance as they represented harmony between humans and nature.

It was in the Tang Dynasty (618-907 CE) when penzai reached new levels of fame in China. The Chinese practice of cultivating miniature trees began to be a source of inspiration for other cultures such as Japan as well as Vietnam.

Japan particularly Japan, in particular, has embraced Bonsai with great reverence. The

Japanese phrase "Bonsai" emerged from the Chinese characters, however the meaning of it changed. It was in Japan, Bonsai came to be a symbol of not just the process of cultivating tiny trees, as well as the technique of shaping them in order to replicate the beauty of large trees. This change was the catalyst for the creation of Bonsai art form that you see it today.

The Kamakura the Kamakura period (1185-1333 CE) in Japan, Bonsai gained immense popularity in the samurai elite class. They considered it a symbol of their connection with the natural world and a reflection of their status as a social being. In the course of time, Bonsai continued to evolve through different styles of thought and styles of regional origin developing.

At the beginning of the 20th century Bonsai attracted the attention of the Western world thanks to the exhibits that were held in international expositions. The exposure resulted in the worldwide spread of Bonsai as an art that attracted enthusiasts and artists all over the world.

In the present, Bonsai stands as a symbol of the everlasting combination of creativity in art and the resilience of nature. The Bonsai's rich and varied history is the profound experience of spanning the globe, cultures, and ages, providing the chance to see the exciting connection between humanity as well as the natural world.

The Philosophy of Bonsai

Bonsai isn't just the result of a gardening venture, it's an incredibly philosophical concept which embodies a profound respect for balance, nature as well as the passing of the passage of. The ancient art of bonsai is deeply rooted in a thought that goes far beyond basic cultivating miniature trees. it's an entanglement between humans and the natural world.

1. Simple and minimal The underlying principle of Bonsai principles is the idea of

simplicity. Bonsai enthusiasts seek to capture the essence of trees with the simplest form possible. That means focusing on the most essential branches and leaves by removing all distractions in order in order to expose the tree's true essence. This is how Bonsai is a perfect example of the Japanese aesthetic concept in the form of "Wabi-sabi," which celebrates imperfections and the transient.

2. Harmony and Nature: Bonsai is a dialog between the artists and trees. Artists must be aware of the nature of the tree's tendencies, pattern of growth, as well as its needs. By understanding this the Bonsai artist can guide the development of the tree, and then shapes the tree while respecting its natural nature. The goal is to design an appearance of a tree as like it would be within the natural world, and harmonises with the environment around it.

3. Timing and Patience: Bonsai exemplifies the patience that is required for both art as well as in everyday life. The miniature trees

typically require a long time, even years, to attain their ideal shape. People who are trained to recognize the steady, slow progression of time. It is similar to the growth of the tree it self.

4. Traditionality and continuity: Bonsai carries traditions passed over generations. Bonsai practitioners respect the tradition of their predecessors taking into consideration established techniques and designs, while also allowing to personal expression. This connects both the present and the past, enhancing the feeling of timelessness Bonsai is a symbol of.

5. Balance and Proportion Bonsai is a rigorous exercise to the balance and proportion. Every aspect of the tree, from the dimensions and position of the branches, to the form of the trunk and the shape of the trunk, is assessed. The pursuit of balance is a reflection of the larger philosophy that seeks harmony in everything we do.

6. Resilience and Adaptation: Bonsai trees, though smaller in height but are incredibly resilient. They can withstand harsh circumstances, a symbol of our human capacity to adapt and prosper when faced with adversity. Bonsai is a reminder that beauty can come out of adversity like a carefully-crafted tree emerges from thoughtful cultivating.

7. Meditation and mindfulness Many practitioners find that practicing Bonsai is a contemplative and mindful exercise. This allows them to completely become completely present in the moment and focus their attention completely on the tree and the process of shaping it. This kind of mindfulness creates a profound relationship with the trees and the environment that surrounds the tree.

The underlying philosophy of Bonsai goes beyond the act of cultivation itself. The practice encourages the practitioner to think about the nature of life's fleeting moments as well as the wonder of imperfections as well as the significance of harmony and balance. Bonsai serves as a powerful reminder that through the process of making trees, we can influence our understanding of the world as well as the place we play in it.

10 BEST BONSAI TO START WITH FOR BEGINNERS

This is a short introduction to each of the top 10 bonsai trees to start with, as well as their primary qualities:

1. Ficus Bonsai (Indoor)

Specifications: Ficus Bonsai, scientifically identified as Ficus Retusa, is one of the most sought-after choices for bonsai lovers who live indoors. It is renowned for its shiny, extremely elliptical leaves, and its stunning aerial roots, Ficus is an exotic plant indigenous in Southeast Asia. It flourishes in warmer, more stable climates typically encountered indoors. It also responds to regular trimming, making it an ideal option for those who are just starting out.

2. Cork Bark Chinese Elm (Versatile, Indoor or Outdoor)

Features: Cork Bark Chinese Elm (Ulmus parvifolia) is a pliable bonsai tree that is able to be planted both outdoors and inside which makes it a good fit for different conditions. It's prized due to its corky bark, and its tiny deep green leaves. This plant is renowned as being tough, making it the perfect option for beginners to bonsai.

3. The Jade (Indoor)

Specific Features: Jade Bonsai (Crassula ovata) is a succulent plant that is indigenous to South Africa. It is characterized by its fleshy, oval leaves, and broad stems that store the water. It is ideal to cultivation in indoor spaces and requires little maintenance. The unique design and low demands on water makes it a perfect option for those who are just starting out.

4. Fukien Tea (or Carmona) (Indoor)

Features: Fukien Tea (Carmona microphylla) is a tiny evergreen tree that has small dark green leaves as well as small white flowers. Fukien Tea is an extremely well-known indoor bonsai because of its small size and beautiful leaves. Fukien Tea is a difficult plant to nurture, however it can be enjoyable for those who like working on the bonsai techniques they have developed.

5. Sweet Plum (Indoor)

Specifications This Sweet Plum (Sageretia theezans) is a deciduous shrub with small, delicate leaves as well as elegant branches. It is a small tree that produces fragrant white flowers as well as tiny red fruit. It is a great choice to indoor bonsai cultivation and is highly regarded because of its beauty and its suitability to bonsai styling.

6. Juniper Bonsai (Outdoor)

"Juniper Bonsai" (Juniperus) is an well-known option for bonsai lovers who enjoy outdoor use. It is comprised of a range of species that are renowned by their rough look, needle-like foliage and distinct bark. Junipers are tough and can withstand all environmental conditions, making an excellent alternative for people who want to start to enjoy bonsai in the outdoors.

7. Cotoneaster (Outdoor)

The Cotoneaster species are bonsai plants that can be found in the outdoors. They're that are renowned for their petite oval-shaped leaves as well as beautiful, sprawling pattern of growth. They are able to produce beautiful flowers of pink or white in spring and red-colored autumn berries which adds to their aesthetic attractiveness. Cotoneasters are tough and can adapt very well to conditions in the outdoors.

8. Japanese Maple (Outdoor)

Specifications: Japanese Maple (Acer palmatum) is known for its beautiful deep lobed leaves which display vibrant autumn colors. It's a perfect option for bonsai growing outdoors that provides all year round visual appeal. Japanese maples flourish in partial shade and demand meticulous attention to details when it comes to their care.

9. Azalea (Outdoor)

Features: Azalea bonsai (Rhododendron indicum) is renowned by its multitude of bright trumpet-shaped spring flowers. Bonsai in the outdoors are loved for their gorgeous blooms as well as the lush green leaves. Azaleas require special attention for their healthy appearance and vitality.

Chapter 2: Ficus Bonsai Care And Maintenance

Origin of the Ficus Bonsai, scientifically known as Ficus Retusa, has its roots to subtropical and tropical areas in Southeast Asia. This adaptable tree belongs the Moraceae family and has grown to become the most sought-after choice of bonsai fans due to its beautiful characteristics and ease of taking care of.

Indoor Bonsai Ficus Bonsai is predominantly cultivated for indoor use. Its adaptability to indoor environments, along with its attractive appearance, is a popular alternative for those interested in bonsai who do not possess access to outside space to grow. If it is grown indoors it is thriving in stable warmer, comfortable environments that have well-controlled lighting and humidity levels.

Specific requirements regarding temperature and humidity Maintaining a suitable level of humidity is crucial to ensure the wellbeing of Ficus Bonsai. It is recommended that it be maintained in an area that has moderate or high levels of humidity. It can also be adapted to lower levels of humidity inside. When it comes to temperatures, Ficus prefers warm conditions ideal between 65degF and 75degF (18degC up to 24degC). It is safe from drafts and extreme temperature swings.

Affordable Insolation Ficus Bonsai thrives in bright indirect light. Set it next to a window that is filtered, or utilize grow lights to give it the needed light. Keep it away from direct sun that could burn the leaves.

Step-by-Step Growth Guide for Ficus Bonsai:

Choose the container: Pick an elongated, well-drained bonsai container that permits the growth of roots.

Type of Soil: Make use of the soil mix that drains quickly usually made of constituents such as pumice, akadama and the lava rock.

Watering guidelines: Make sure to keep the soil always moist. However, be careful not to overwater, since Ficus Bonsai is susceptible to root decay. Make sure to water thoroughly when you notice that the top inch of soil is feeling dry.

Root structure pruning Prune the roots when the process of repotting. This usually happens every 2 to 3 years. Prune the roots of any excess to ensure the health of the tree and to make room for its pot.

Fertilize your soil: Feed your tree once every 4 to 6 weeks throughout the growing season (spring through the summer) by using a balanced water-soluble fertilizer that contains micro and macronutrients.

Best Placement: Put the Ficus Bonsai in a location that has indirect sunlight. You can also utilize artificial grow lights to ensure adequate lighting.

Further Pruning and wiring: Prune to form the bonsai in its growth, and trim off new growth in order to preserve its desired shape. Wire is a great tool to shape the bonsai, however Ficus is surprisingly adaptable and will require lesser wiring than some species.

Ficus Bonsai, with its shiny leaves and exquisite aerial roots, provides an impressive space for lovers of bonsai. The durability of the plant, coupled with simple care instructions make it a great option for novices and experienced experts alike. With the right environment with regular care the plant will grow into an incredibly beautiful and flourishing Ficus Bonsai.

CORK BARK CHINESE ELM CARE AND MAINTENANCE

Origin: Cork Bark Chinese Elm, also known scientifically as Ulmus parvifolia, is indigenous to China, Korea, and Japan. Its versatility and distinct bark texture has become a top option for bonsai lovers from all over the world.

Flexible (Indoor and outdoor) Versatile (Indoor or Outdoor): One of the unique characteristics that is unique to Cork Bark Chinese Elm is its versatility. Cork Bark Chinese Elm is the versatility. It's a great choice for an outdoor and indoor bonsai which makes it an ideal choice for a variety of environments and climates.

Humidity and Temperature Requirements:

Humidity: Chinese Cork Elms like moderately humid levels. In order to increase the humidity, put a humidifier tray with pebbles, water and the bonsai.

The bonsai has the ability to be adapted to various temperatures and prefers mild temperatures. It's able to withstand cool and hot temperatures. This makes it an ideal option for a variety of climates.

Affordable Insolation: Cork Bark Chinese Elm enjoys the full shade to some shade. In the outdoors, it is able to benefit from the direct sun for an hour or two. If you are growing it indoors, be sure it is exposed to bright and indirect sunlight. The bonsai should be rotated regularly for a consistent growth.

Step-by-Step Growth Guide for Cork Bark Chinese Elm:

Selection of the Container: Select an appropriate bonsai planter that can accommodate the root system, and gives the space needed to grow. The container should be equipped with drainage holes.

Soil Type: Use an aeration-friendly bonsai soil mix that typically includes elements like pumice, akadama and lava rocks.

Watering guidelines: Keep your soil always moist but do not let it become waterlogged. Let the top inch of soil to dry a bit before repeating the watering.

Pruning the Roots Structure Prune the roots when Repotting, usually repeated every 2 to 3 years. Cut back roots that are too long in order to keep the bonsai healthy and ensure it fits into the pot.

Apply fertilizer to Your Cork Bark Chinese Elm every 2 to 4 weeks throughout the period of growth (spring until early autumn) by using a balanced water-soluble fertilizer.

The best position is if you plant it outdoors, put it in the direct light for a portion all day. If indoors, provide bright, indirect light near a window.

Other wires and pruning: Regularly pruning is necessary to preserve the ideal form and dimension of your bonsai. Wiring is used to direct the branch and the trunk, but Cork Bark Chinese Elm is a different species. Cork Bark Chinese Elm tends to be a flexible tree and could require less wire than different species.

The care of the Cork Bark Chinese Elm bonsai allows you to experience its versatility and distinctive bark texture, while enhancing the skills of cultivating bonsai. No matter if you want to cultivate the plant outdoors or indoors with the proper conditions and proper attention can ensure a healthy, attractive bonsai.

THE JADE CARE AND MAINTENANCE

Origin: Jade Bonsai, scientifically known as Crassula Ovata, is native to desert landscapes in South Africa. The unique look of the plant, which is characterized by its fleshy oval-shaped leaves as well as strong, robust stems, have given it a position in the field of bonsai.

Indoor Bonsai Indoor Bonsai Jade Bonsai is predominantly cultivated in an indoor setting. Its capacity to flourish in indoor conditions with minimum effort makes it an ideal option for those who prefer low maintenance plants.

Humidity and Temperature Requirements:

Humidity Jades are succulents, and therefore, they can be tolerant of low humidity levels that are typically found in indoor spaces. But, they can appreciate occasional misting in order to improve the humidity.

Temperatures: Jades like cool indoor temperatures. They prefer that range between 65degF and 75degF (18degC up to 24degC). They can be sensitive to cold breezes and must be shielded from abrupt temperature drop.

Affordable Insolation: Jade Bonsai thrives in bright indirect light. Set it in an east-facing or south-facing windows where it will receive the sun's rays in a controlled manner. Beware of exposing it to harsh bright sunlight as it can result in sunburn to the leaves.

Step-by-Step Growth Guide for The Jade:

Choose the right container: Select the bonsai pot that is shallow and has adequate drainage. The pot must be designed so that it can accommodate the root system, and also ensure the stability.

Soil Type: Utilize a well-draining bonsai soil mix. Soil blends made of succulents or cactus tend to be suitable for Jades.

Watering Tips Let the soil dry slightly in between the watering. Use sparingly, and try to avoid overwatering because Jades are very vulnerable to root decay.

Root structure pruning Pruning roots of Jade Bonsai is generally infrequent. Repotting each 2-3 years could be accompanied by minor root pruning principally to refresh the pot and soil.

Fertilization of the soil: Apply fertilizer moderately throughout the growing period (spring until early autumn) by using a balanced dispersed liquid fertilizer. Limit or stop fertilization in winter.

Ideal Position: Place Your Jade Bonsai near a sunny windows that receives bright indirect sunlight. If the natural light you receive is not enough then you should consider adding growing lights to complement.

Additional Pruning and Wiring is vital to keep the shape and size you want the Jade Bonsai. Wiring is seldom required because of the natural pattern of growth, but it is sometimes used to make minor adjustments to the shape of your tree.

Chapter 3: Fukien Tea Care And Maintenance

Origin: Fukien Tea, scientifically known as Carmona Retusa, is a native of the south regions in China, India, and Southeast Asia. This evergreen, tropical plant has grown to become an increasingly popular option for those who love bonsai because of its tiny shiny leaves, smooth and beautiful white blooms.

Indoor Bonsai Fukien Tea is primarily cultivated for indoor bonsai. The preference for warm and sturdy conditions works well in indoor settings, which makes it the perfect alternative for those looking to keep bonsai in indoor environments.

Humidity and Temperature Requirements:

Humidity Fukien Tea thrives in areas with moderate to high levels of humidity. A regular misting program or trays of humidity are a great way to ensure adequate humidity levels in the bonsai.

Temperature: The species is a fan of warm temperatures. Ideally, it should be between 65degF and 75degF (18degC up to 24degC). Be sure to protect it from air and extreme temperature swings.

Affordable Insolation: Fukien Tea benefits from direct, bright sunlight. Set it in a location which is well-lit, as direct sunlight may scorch the leaves. If the natural light source is not enough then add the use of grow lights to provide adequate lighting.

Step-by-Step Growth Guide for Fukien Tea:

Choose a container: Go to use a deep, well-draining bonsai planter that is able to allow for the growth of roots. Be sure to check that the pot is equipped with drainage holes.

Soil Type: Choose an aeration-friendly bonsai soil mix which typically contains components such as pumice, akadama and the lava rock.

Watering guidelines: Keep your soil moist and not overly waterlogged. Be sure to water the soil thoroughly once the soil's top inch appears dry. Don't let the soil become completely dry.

Root Pruning Prune the roots prior to the process of repotting. This is usually performed every two to three years. Cut off any roots that are not needed to keep the bonsai healthy and allow it to fit into its.

You should fertilize the soil around your Fukien Tea every 4-6 weeks in the season of growth (spring to summer) using a balanced, water-soluble fertilizer.

The ideal position: place Fukien Tea Fukien Tea near a window which has indirect light, or utilize artificial grow lighting to give adequate illumination.

Further Pruning and Wiring Pruning is essential to form the bonsai, and keep it in a compact shape. Wire is a way to direct the branches, however Fukien Tea's branche are adaptable and so only a small amount of wiring is usually required.

Growing an Fukien Tea bonsai offers an possibility to be able to see the beautiful nature of this species as well as learn to understand the special care requirements for bonsai growth in the indoor environment. When you pay the appropriate treatment and care to your Fukien Tea bonsai can thrive and grow indoors.

SWEET PLUM CARE AND MAINTENANCE

Origin of the Sweet Plum, scientifically known as Sageretia theezans is found in subtropical areas of Asia that includes China as well as Malaysia. The tree is admired by its tiny, beautiful leaves, and beautiful flowers has been recognized as a beautiful bonsai tree.

Indoor Bonsai Sweet Plum is often cultivated for indoor use and is appreciated for its flexibility to conditions in the indoor. Its beautiful design and small size makes it a wonderful option for those who love bonsai.

Humidity and Temperature Requirements:

Humidity Sweet Plum excels in conditions of moderate to high humidity. A regular misting routine or putting the bonsai inside an airtight tray will aid in maintaining the proper moisture levels.

Temperature: The species likes mild temperatures, best with a range of 65-75degF (18degC up to 24degC). Beware of exposing the bonsai to cold winds or sudden temperature changes.

Affordable Insolation Sweet Plum is a great choice for direct, bright sunlight. It should be placed near the west or south windows where it will receive the sunlight in a way that is filtered. If light from the sun isn't enough you can add the grow light to give it adequate lighting.

Step-by-Step Growth Guide for Sweet Plum:

Choose a container: Pick the bonsai pot that is shallow and has adequate drainage. The pot must be large enough to support the root system as well as give security.

Type of Soil: Choose an aeration-friendly bonsai soil mix typically made up of elements such as pumice, akadama and the lava rock.

Watering Tips: Maintain an even amount of moisture in the soil throughout the day, and avoid the effects of drought as well as the logging of water. Make sure to water thoroughly every time the top of your soil appears dry.

Pruning the Roots Structure Prune the roots when the process of repotting. This is usually done every two to three years. Pruning the roots back in order to maintain the bonsai's condition and to fit it into the pot.

Fertilize your sweet Plum every 4 to 6 weeks during the period of growth (spring until early autumn) by using a balanced water-soluble fertilizer.

Best Place: Put the Sweet Plum in a location near windows that have indirect light, or you can use artificial grow lights to guarantee enough lighting.

Other Pruning and Wiring: Keeping the pruning schedule regular is necessary to shape the bonsai, and keep its elegant appearance. Wiring is a great option to make structural adjustments. However, Sweet Plum branches are flexible and often require only the least amount of wiring.

The care of a Sweet Plum bonsai offers an chance to experience the beauty and grace of this plant while also gaining knowledge of growing bonsai. If you take care of it properly and pay attention the Sweet Plum bonsai can flourish in the indoor environment, delighting both you and those who visit.

JUNIPER BONSAI CARE AND MAINTENANCE

The origin of the species: Juniper Bonsai belongs to the Juniperus Genus that has an extensive global spread. The hardy tree are found in a variety of environments, ranging from the mountain ranges that are found in North America to the Mediterranean regions, and even further. Juniper Bonsai varieties have been grown for centuries, which makes them a popular selection for those who love bonsai in the outdoors.

Outside Bonsai: Juniper Bonsai is typically an outdoor bonsai plant. The ability to adjust to outdoor conditions such as temperature variations and sunlight from the sun, makes it a great option for people looking to experience an experience that is truly bonsai-like.

Humidity and Temperature Requirements:

Humidity: Juniper bonsai is well-suited to humid areas. Its toughness allows it to flourish in dry, humid air.

Temperature: The Juniper have the ability to stand up to a range of temperatures. They're extremely cold-resistant and are able to withstand frost as well as cold winters. This makes them the ideal plant for mild conditions.

Affordable Insolation: Juniper Bonsai thrives in the full shade to the partial shade. When it's grown outdoors the plant should be exposed to the sun's direct light for some hours per day. A proper mix of sunshine is essential to the health of your plant and its growth.

Step-by-Step Growth Guide for Juniper Bonsai:

Choose a container: Select the bonsai's well-drained, shallow pot that is in harmony with the bonsai's style and size. Be sure to use a pot with drainage holes in order to avoid waterlogged soil.

Type of Soil: Choose an aeration-friendly bonsai soil mix with components that include pumice, akadama, and the lava rock.

Watering Tips Apply water thoroughly when the soil's top layer is feeling dry. The Juniper plants prefer a constant water level, but they should not be left in water that is stagnant. Alter the frequency of irrigation according to the climate and time of year.

Pruning the root structure Juniper Bonsai does well with root pruning when repotting is done, which typically occurs each year for a period of 2 or 3 years. Cut back roots that are too long in order to keep the tree healthy and to make room for the pot.

Fertilize your Juniper Bonsai each 4 to 6 weeks throughout the growing season (spring to summer) by using a balanced slow-release fertilizer.

Best Placement: Put the Juniper Bonsai in an outdoor area that gets plenty of sunlight. Be sure to shield it from strong storms, because the force of winds could dry the leaves and cause damage to branches.

Other Pruning and Wiring Juniper Bonsai needs regular pruning in order to form the shape of its leaves and branches. Wiring can help provide guidance to branches, encourage movements, and help achieve the desired appearances.

The care of a Juniper Bonsai offers the chance to witness the natural beauty of these trees, while improving your bonsai knowledge.

When you take care of it properly and in an outdoor space that is suitable Your Juniper Bonsai can flourish and grow into its unique windswept style as time passes.

Chapter 4: Cotoneaster Care And Maintenance

The origin of the species: Cotoneaster is a genus of flowering plants indigenous to a variety of areas, which includes Asia, Europe, and North Africa. These robust and adaptable plants have been brought into the bonsai world with distinctive characteristics as well as ornamental value.

Outdoor Bonsai Cotoneaster is typically cultivated for an outside bonsai. Its inherent resilience to the elements, along with its capacity to endure temperatures that fluctuate, make it an ideal choice for lovers of bonsai who want to enjoy the outdoors experience.

Humidity and Temperature Requirements:

Humidity: The Cotoneaster Bonsai can be adapted to varying humidity levels, and is able to endure moderate to low levels of humidity. It's a great choice for dry conditions.

Temperature: The species is typically robust and is able to withstand the wide range of temperatures. It is able to thrive both warm summers as well as cold winters so it is a good choice for diverse environments.

The right insolation for the plant: Cotoneaster Bonsai appreciates full sun and partial shade. Set it up in an area in which it will receive full sun during a part of the time. The proper amount of sunshine is crucial to ensure healthy growth and blooming of the berries and flowers.

Step-by-Step Growth Guide for Cotoneaster:

Selection of the Container: Pick the bonsai pot that is shallow and has the proper drainage. Make sure the size of the pot is suitable for the root system, and matches the design you'd like to see in the Cotoneaster Bonsai.

Soil Type: Use an aeration-friendly bonsai soil mix with components that include pumice, akadama and the lava rock.

The guidelines for watering: keep your soil moist and be careful not to overwater. Let the top layer of soil to slightly dry prior to watering it again. Alter the frequency of irrigation according to the climate and time of year.

Pruning the Root Structure Pruning the roots is usually done when repotting takes place every 2 to 3 years. Pruning the roots of a tree is a good way to ensure the health of the tree and allow the pot to be affixed.

Feed your Cotoneaster Bonsai each week for 4-6 weeks during the growth season (spring through the summer) using a well-balanced, water-soluble fertilizer for your bonsai.

Best Placement: Put it Cotoneaster Bonsai in an outdoor spot that is well-lit and fresh airflow. It should be protected from winds that are strong and storms, which could damage leaves and branches.

Other pruning and wiring: Pruning is necessary to shape the Cotoneaster and to encourage the ramification process. Wiring is used to make stylistic adjustments or structural changes but is best done with caution since Cotoneaster branches are brittle.

The care of a Cotoneaster bonsai allows you to experience the beauty and natural appeal of these tolerant shrubs as well as improving your bonsai-growing skills. If you take care of it and pay focus on environmental conditions Your Cotoneaster Bonsai can thrive and showcase its beautiful blooms throughout the year and the berries.

JAPANESE MAPLE CARE AND MAINTENANCE

Origin: Japanese Maple, known as Acer palmatum is native to Japan, China, and Korea. The tree is admired for its beautiful leaves and breathtaking seasonal changes This tree has captured bonsai fans from around the globe.

The Outdoor Bonsai: Japanese Maple is predominantly grown for outdoor bonsai. The need to have an extended winter period of dormancy and its ability to adapt to indoor environments makes it ideal to grow outdoors.

Humidity and Temperature Requirements:

Humidity Japanese Maple Bonsai favors moderate to high levels of humidity and can achieve this via regular misting, or placing the bonsai onto the humid tray.

Temperature: The tree flourishes in warm climates. Japanese Maples are tolerant of the cold winters as well as mild summers. Temperatures range from 45degF-85degF (7degC up to 29degC).

Affordable Insolation Japanese Maple Bonsai benefits from partially shaded or dappled light. It is best to protect it from intense afternoon sunlight, particularly during the hot summer months as it may scorch delicate leaves. The morning sun is usually tolerated.

Step-by-Step Growth Guide for Japanese Maple:

Selection of the Container: Select an unassuming bonsai container with adequate drainage. The pot must complement the design of the tree, and also provide space for the growth of roots.

Soil Type Choose the bonsai soil mix that is well-drained usually includes elements like pumice, akadama and the lava rock.

Watering guidelines: Keep the soil always moist, but not overly waterlogged. Be sure to water well when the soil's top layer is feeling dry. Change the frequency of irrigation depending on the weather and the season.

Pruning the Roots Structure The roots are pruned during the process of repotting. This is usually carried out every 2 to 3 years. Cut back roots with excess to ensure the health of the tree and allow the pot to be affixed.

You should fertilize the soil around your Japanese Maple Bonsai every 4-6 weeks in the growing period (spring to summer) with a well-balanced Bonsai fertilizer that is water-soluble.

The best position is to place it Japanese Maple Bonsai in a area that is filtered by sunshine or some shade. Be sure to shield it from winds that are strong and storms, which could damage the delicate leaves.

Additional pruning and wiring: Pruning is necessary to shape the bonsai's shape and improve its branching pattern. Wire is a great option for styling purposes, but you should do it cautiously, because Japanese Maple branches can be hard and brittle.

Maintaining Japanese Maple Bonsai provides you the chance to experience the changing beauty of this tree throughout the year. When properly cared for and attentive to the needs of its environment Your Japanese Maple Bonsai can become an amazing part of your collection of bonsai.

AZALEA CARE AND MAINTENANCE

Source: Azaleas are part of the Rhododendron genus originate from a variety of regions which include Asia, Europe, and North America. They are stunning flowering plants that have been brought into the bonsai world, with vibrant flowers as well as lush greenery.

Outdoor Bonsai Azalea Bonsai is mostly an outdoor bonsai plant. The time of dormancy during the season and their love for outdoor environments that are natural make ideal for growing outdoors.

Humidity and Temperature Requirements:

Humidity Azalea Bonsai flourishes in conditions of high humidity. In order to maintain a healthy level of moisture frequently misting or using the humidity tray can be beneficial.

Temperature: favor mild climates that have cool winters as well as warm and humid summers. Ideal temperatures are between 45degF-85degF (7degC up to 29degC).

Affordable Insolation Azalea Bonsai prefers dappled sun or some shade. Keep them away from direct, strong sunlight during the hottest times of the day since it may damage delicate flowers and leaves. In general, morning sunlight is well-handled.

Step-by-Step Growth Guide for Azalea:

Choose a container: Select an un-shallow bonsai container with great drainage. The container should fit in with the style of the bonsai and provide room for root growth.

Soil type: Use an acidic and well-drained bonsai soil blend specifically formulated specifically for the azalea. The mix typically includes elements such as peat moss, perlite and pine bark.

Watering guidelines: Keep the soil always moist, but not soaking wet. Make sure to water thoroughly whenever the soil's top layer is feeling dry. Change the frequency of irrigation depending on the weather and season.

Pruning the Root Structure Pruning the roots is usually done in repotting. It occurs every two to three years. Cut back roots that are too long to ensure the bonsai's overall health and to make room for the pot.

Make sure to fertilize the soil around your Azalea Bonsai every 4 to 6 weeks throughout the growing season (spring through the summer) using a well-balanced, water-soluble fertilizer specifically designed to plants that love acidity.

Best Place: Put your Azalea Bonsai in an area that has dappled sun or shade. Protect it from windy conditions that can harm the foliage and flowers.

Chapter 5: Pines Care And Maintenance

Origin: The pine trees form part of the Pinus Genus that has the world's largest spread. They can be found across many regions including North America, Europe, Asia and many more. Their distinctive appearance and long-lasting life make them an extremely popular selection for bonsai fans.

Outdoor Bonsai The majority of pines are grown as bonsai for outdoor use. They need outdoor conditions in order for growth, which includes exposure to the seasonal fluctuations in temperature.

Humidity and Temperature Requirements:

Humidity: They generally can tolerate a variety of levels of humidity. They can be adapted to humid and dry climates.

Temperatures: Pines are cold-hardy and flourish in frigid winters. They are tolerant of distinct seasonal variations and endure temperatures from -10degF up to 90degF (-23degC up to 32degC).

Affordable Insolation: Pines require full sun for their growth. Put them in a place that receives full sunlight for most all day. A good amount of sunlight is vital for healthy growth as well as the establishment of dense vegetation.

Step-by-Step Growth Guide for Pines:

Selection of the Container: Select the bonsai planter with a great drainage capacity to hold the pine's deep taproot trees. The planter should be in harmony with the design of the tree, and also provide enough space to allow for root growth.

Soil Type: Choose the bonsai soil that is well-drained specially designed to work with pines. It typically contains components like pumice, akadama along with pine bark.

Watering guidelines Let the soil somewhat dry between the watering. Make sure you water the soil thoroughly and avoid logging. Change the frequency of watering based on weather conditions and the time of the year.

Root structure pruning Pruning the roots is essential when repotting generally occurs every 2 to 3 years. Cut back roots that are too long in order to keep the tree healthy and to make room for the pot.

Make sure to fertilize the soil around your Pine Bonsai every 4-6 weeks throughout the growing season (spring to summer) by using a balanced slow-release fertilizer.

Best Placement: Position your Pine Bonsai in full sun in a location that receives the most direct light throughout the time. You should ensure that the airflow is good to avoid the growth of fungal diseases.

Other Pruning and Wiring Regular pruning is vital for maintaining the ideal shape as well as to encourage back-budding and decrease the length of needles. Wiring can help form the branches, and to aid in the growth of branches.

Care for your Pine Bonsai offers the opportunity to admire the elegant beauty of these trees that live for a long time with a small size. When properly cared for and attentive to the outdoor environment the Pine Bonsai can thrive and make a beautiful part of the bonsai collection.

MATERIALS, TOOLS, AND EQUIPMENT

Making and maintaining bonsai plants will require the use of a range of tools and equipment that will ensure that you take care of and develop of the miniature trees. This is a checklist of essential equipment required for bonsai.

1. Pruning Shears (Bonsai Scissors):

The Pruning Shears also known as bonsai shears, are specifically designed to ensure exact trimming and cutting leaves, branches, as well as small roots.

Utilize them to create a tree shape and remove any unwanted growth and keep the bonsai's overall design.

2. Concave Cutters:

o Use: Concave cutters possess an angled cutting edge which produces concave cuts when cutting branches. They heal faster than cuts that are flat.

Usage: Ideal for cleanly removing large branches and minimising the risk of scarring.

3. Wire Cutters:

Wire cutters can efficiently cut bonsai wire, without causing damage to the trees.

Their purpose is to take wire off or re-adjust it which was used to create a shape or form branches.

4. Branch Bender (Bonsai Jack):

O Function: A bonsai jack or branch bender can gently shape and bend branches with no risk of damaging.

Utilization allows you to alter branches to create the bonsai style you want.

5. Knob Cutters:

Knob cutters feature the advantage of having a concave cutting blade and are useful in taking out protrusions, or knobs that remain after the removal of branches.

Uses: Speeds up healing, and also creates an even surface following trimming.

6. Root Hook or Rake:

O Function: A root hook, also known as a rake, is utilized for untangling the roots, as well as combing the roots when the process of repotting.

Helps prepare the root ball to be pruned and the repotting.

7. Root Pruning Shears:

The function of the root pruning shears were designed to be used to trim and prune the roots of bonsai plants.

Use: Essential for repotting, to preserve the health of the root system and adapt to the bonsai planter.

8. Bonsai Wire:

The function of aluminum or copper bonsai wire can be utilized to form and form branches through offering the necessary support and direction.

Use: wrapping branches around them and then bending them to the position you want them to be.

9. Bonsai Soil Mix:

The Use: Bonsai soil mixtures are created to ensure adequate drainage and aeration, and also to retain moisture.

Use: As the bonsai's growing medium trees, to help promote the development of healthy roots.

10. Bonsai Pots:

The purpose of the bonsai is to create shallow pots with drain holes are chosen to hold the bonsai and enhance its design.

Useful Your container is the one you use that you use for your bonsai has been selected in accordance with design and functionality.

11. Chopsticks or Wooden Sticks:

The function is to perform different tasks, such as the removal of air pockets during repotting, and placing soil in the potting.

Utilization: Facilitates exact soil positioning and the removal from air pockets.

12. Bonsai Turntable:

O Function: A turning table makes it easy to rotate the bonsai to ensure all-round maintenance cutting, trimming and styling.

Utilization: enables access to different perspectives of the tree while not moving it.

13. Watering Can or Nozzle:

The purpose of a watering can or nozzle gives the water in a soft stream to ensure that watering is controlled and does not disturb the soil.

Use: Provides sufficient irrigation while stopping soil erosion and disturbance to the roots.

14. Bonsai Brush or Toothbrush:

O Function: Soft brush or brushes are employed to scrub the bonsai's branches, trunks and leaves.

Useful: Get rid of dirt, dust, and algae off the bonsai's surface in order to preserve the appearance.

15. Wire Brushes:

A Function: Wire brush can be used to clean instruments and to remove algae or moss from bonsai's.

Use: Maintains the equipment in top condition and preserves the bonsai's aesthetic.

16. Bonsai Fertilizer:

The purpose of bonsai is to fertilizers are essential for good growth and development.

Use: Apply frequently during the growing seasons in order to meet the tree's nutritional demands.

17. Screen or Mesh:

A Function: Mesh screens or mesh made of plastic are put over the drainage holes in pots to keep soil from flowing out of.

Use: Aids in maintaining an appropriate drainage, while also retaining soil in the.

18. Misting Bottle:

O Function: A misting bottle helps keep the humidity level of indoor bonsai trees.

Use: Creates the bonsai with a fine mist in order to make a humid environment for bonsai.

Chapter 6: Buy Or Create From Scratch?

The decision of buying a bonsai from a store or making one yourself is based upon a number of variables, such as the level of your knowledge as well as your time commitment and individual preferences. These are the benefits and cons of each for you to make an informed decision:

Buying a Bonsai:

Advantages:

1. Instant Gratification: You receive an all-grown and well-styled bonsai in a matter of minutes, allowing you to take pleasure in the beauty of it right away.

2. For beginners If you're new to bonsai, purchasing a trained tree will make it less scary as well as a great start to learning about how to care for your bonsai.

3. Access to established Trees The trees you can visit are rare or mature bonsai varieties which could take some time to establish completely from beginning to finish.

Disadvantages:

1. Cost: Bonsai trees that are established may be costly, especially in the case of older trees or have exceptional quality.

2. Limited customization: The tree's style is established, which limits the amount of creativity you can add.

3. Maintenance requirements: Even purchased bonsai plants require regular upkeep and attention to ensure their health and flourishing.

Create the Bonsai by scratch:

Advantages:

1. Complete Creative Control: Beginning at the beginning allows you to form and design your bonsai in accordance with your own vision.

2. Learn Experience: You'll develop important knowledge and experience regarding pruning, horticulture and even design.

3. Cost-savings: Creating the bonsai using seeds or young plant could save money in the end.

Disadvantages:

1. A lot of patience is required: Growing bonsai trees from scratch takes patience and time. It can take for several years or more.

2. The Learning Curve: If you're beginner to bonsai it is a long learning curve. Also, mistakes can happen.

3. Initial investment: Although it may be economical over the long run however, it is a first purchase of tools, equipment and plants that are still young.

Considerations:

1. Expertise Level: If you're an aspiring beginner, beginning using a purchased bonsai or a trained tree could provide a great start to learn about this hobby. When you're comfortable then you'll be able to move on to making your own bonsai from scratch.

2. Time Commitment: Determine your time and how long you'll be able to spend on bonsai maintenance. If you are short on time, purchasing an existing bonsai could be the best option.

3. Budget: Determine your budget for bonsai. Don't just think about your initial investment, but also the continuing maintenance expenses.

4. The Creative Vision there is an idea for your bonsai, and you want complete creative control, beginning by scratch is the right option.

5. Learner's Goals: If you're keen to master about bonsai or the cultivation of horticulture from scratch, this course can be a great learning opportunity.

There's not a universal solution. A lot of bonsai lovers prefer the combination of two methods, such as buying a few trees, and then growing them entirely from beginning from scratch. No matter which method you pick you should take pleasure in the process as well as the aesthetics of your bonsai's when it grows over time.

What is the best way to select the bonsai (Climatic space, place and budget)

The right choice for a bonsai tree requires a number of key considerations to determine the health of your tree and its suitability to the specific circumstances and needs of your. How to select one based on the climate of your location, area, and price:

1. Climatic Area:

Be aware of your climate: Learn about the climate that is prevalent in the area you live in, such as the range of temperatures, the levels of humidity as well as seasonal fluctuations. Bonsai trees are characterized by particular climate needs, so choosing one that is compatible with your climate in the area is vital to ensure its health.

Local Species: Search for bonsai species which are native or adapted to your local environment. Indigenous species usually thrive quickly because they've adapted to the local climate. In the case of, say, you live in an extremely dry and hot area look into species that have adapted to the desert, such as Jade or Desert Rose. Jade and Desert Rose.

Microclimates: Check the microclimates that exist in your home or garden. Certain areas may have different climates, so you could choose a bonsai variety to match the microclimate of your area.

2. Placement:

Indoor or outdoor Decide if you want to keep your bonsai in the indoors or outside. It is based on the type of plant you select as well as your climate. Certain bonsai species are suitable for indoor cultivation while others require outdoors conditions.

Sunlight: Take into consideration the intensity and amount of the sun's rays in the area you choose. Bonsai that thrive best in full sunlight, whereas some prefer shade. Select a bonsai species that is compatible with the amount of light that is available.

Protecting yourself from elements Make sure your bonsai's location protects it from harsh weather conditions like intense winds, cold temperatures or excessive hot temperatures. An area that is protected can to maintain the health of your tree.

Area: Bonsai containers come in different sizes. So, choose an appropriate size and tree which is suitable for your space and compliments the decor of your outdoor or indoor space.

3. Budget:

The initial cost: Bonsai trees vary widely in cost, based on the factors such as the species' age, their size, as well as the style. Set a budget for your first purchase. Keep in mind that smaller and more stylish trees can be cheaper.

Costs for ongoing maintenance: Think about the cost of long-term maintenance for your bonsai. These include the tools and soil, fertilizer and any specific care needs of the particular species. Add these costs to your budget.

Investment or. Hobby: Choose which bonsai you see as a way to invest or just as a pastime. Certain collectors prefer to put more money into extraordinary or rare specimens while others are more focused on enjoyment from the art with no any financial investment.

Training vs. established: When you're working on the tightest budget, then you may opt to go with a smaller, more casual bonsai which you can mold and expand with time. Styled and established bonsai may cost more.

Selecting a bonsai which is compatible to your environment, climate as well as your budget will ensure an enjoyable and harmonious bonsai adventure. It is essential to study and speak with knowledgeable bonsai lovers or local nurseries in order to make an educated choice and choose a plant which will flourish under your environment.

The signs of a healthy bonsai tree

Being aware of the indications of a bonsai tree that is healthy is vital for their preservation and care. These are the most important indicators for a bonsai that is healthy:

1. A vibrant foliage Leaves or needles of a bonsai that are healthy should be bright, green, free of discoloration becoming brown or yellowing. The color of the leaves can vary according to the bonsai's species.

2. An appropriate leaf density: A healthy bonsai tree is one that has proper needle or leaf density to its species. The tree should not be excessively packed or lacking.

3. A regular growth pattern: Your bonsai needs to show a consistent, controlled development. The new shoots and leaves must emerge regularly throughout the growing season. The tree should keep its preferred appearance.

4. Strong roots: Bonsai that are healthy have well-developed roots. Take the bonsai out of the container regularly to inspect for healthy, clear, or fibrous, roots. The roots that appear to be tangled and rootbound could indicate the need to repot the bonsai.

5. Sturdy Branch and Trunk The trunk and the branches must be strong and free of indications that indicate decay, rot or any damage. The trunk should slowly reduce in size from the base up to the apex.

6. Healthy Balanced Growth: A well-groomed bonsai will show healthy growth across the entire canopy. Be sure to ensure that one portion of the tree isn't expanding at a rate that is excessively in comparison to other parts of the tree.

7. In absence of any disease or pest Check the bonsai frequently to look for indications of pests or diseases, like abnormal spots, yellowing leaves or infestations of pests. Bonsai trees that are healthy are not as susceptible to such problems.

8. Properly Watered The soil must be equally moist, but not soaked. Correct watering methods assure that the bonsai is receiving enough hydration and does not suffer from dryness or rot in the roots.

9. Regular Pruning: Healthy bonsai plants need regular trimming to keep their form and stimulate the growth of new trees. Pruning should be carried out using safe, sharp tools that are clean to protect against injury.

10. Responds to Care Healthy Bonsai Tree will respond positively to the proper treatment. If it's receiving a sufficient quantity of sunshine along with water and other nutrients, it's likely to prosper and keep growing.

11. Seasonal Changes Bonsai trees are likely to show the seasonal change, like shed of leaves or the development buds. This indicates that the plant reacts to changes in the environment.

12. There are no signs of stress: Bonsai trees in good health are not exhibiting indications

of stress, like wilting leaves as well as excessive yellowing and branch drooping.

13. Sturdy Growth Rings: If you trim a branch take a look at the cross-section. Bonsai trees that are healthy are characterized by distinct and well-spaced growth rings which indicate consistent growth throughout the many years.

14. Soil Health: The soil must be well-aerated and free of indications of waterlogged or compacted soil. Healthy soil promotes healthy roots.

15. Adjusting to seasonal changes Bonsai that are healthy should adjust to the changes in temperature, light and humidity during the season without overstressing.

Careful observation and monitoring as well as a thorough understanding of the needs specific to the bonsai you have chosen, can help maintain an enviable and healthy miniature tree.

How to make the bonsai tree you want to create starting from scratch

Making a bonsai starting from scratch can be a satisfying however time-consuming and patient procedure. This step-by-step tutorial will assist you in the process:

1. Choose Your Bonsai Species:

Choose a plant that is suited to your environment and level of experience. Some of the most common choices for beginners are Ficus, Jade, and Chinese Elm.

2. Acquire a Young Tree:

Get a new sapling or tree from a plant nursery or at a garden centre, or you can propagate by cuttings or seeds. Trees that are younger are more easy to manage.

3. Select the Right Pot:

Select a small bonsai pot that has good drainage. The dimensions and design of the pot must complement the design of the tree in the future.

4. Prepare the Soil Mix:

Make or buy the bonsai mix that is well-drained and that is suitable for the species you want to grow. Most common ingredients are akadama pumice, as well as the lava rock.

5. Repot the Tree:

Remove the baby tree from the container it was in. Cut any tangled or long roots, and then repot the tree into the bonsai planter with fresh soil. Be sure that the tree is evenly placed on the container.

6. Prune the Roots:

Cut back the roots, allowing for an efficient root system. Do not remove more than a third part of the roots mass at one period of. Repot every two to three years to ensure the health of your root.

7. Initial Pruning and Wiring:

Begin to shape the tree by trimming off branches that are not needed and then making the rest of them. Utilize bonsai wire to help guide branches to their desired locations. Take care not to damage the new tree.

8. Allow Growth:

Let the tree develop freely for a time or two, establishing the strength of its trunk and branches. In this time, you must concentrate on maintaining the health of your tree.

9. Prune and Shape Regularly:

Trim and wire the tree on a regular basis in order to keep the shape you want. Take your time; it could be a while before you get the ideal style.

10. Monitoring Health: Maintain a keen eye on the health of your tree, looking for diseases, pests as well as signs of stress. Adjust care accordingly.

11. Repot as needed Repot the bonsai each 2-3 years to replenish the soil and structure of the roots. It also gives you the chance to trim and shape the root.

12. Make sure to fertilize the tree in the right way: Use the right bonsai fertilizer for your needs in the season of growth (spring to summer). The frequency and the strength of fertilization based on the requirements of your bonsai tree.

13. Make sure to water the bonsai once the surface of the soil begins to become dry. The frequency is contingent on variables like the climate, and the size of pot. Be sure to drain the pot properly in order to stop the occurrence of root rot.

14. Be consistent and patient: Bonsai cultivation is a continuous process that demands patience as well as perseverance. The trees can require a few years to develop into an elegant bonsai.

15. Style and refinement: When your tree is mature refine its appearance and its structure. Make use of advanced methods like defoliation, wire, and deadwood styling to create elaborate designs.

The process of creating a bonsai plant starting from scratch is an adventure which combines the horticultural expertise and artistic flair. It's a lengthy commitment which rewards perseverance and commitment and the beautiful look of an ever-growing piece of artwork.

DIY Bonsai Kit

An DIY Bonsai Kit is a practical way for novices to begin their bonsai journey. It usually includes the necessary elements and directions necessary to build and maintain the bonsai tree. This is what you can expect to get inside the DIY Bonsai kit:

1. The Young Bonsai Tree: The kit generally includes a small bonsai, which is often trained at least in part, that is able to be potted and cut to enhance the education journey.

2. Bonsai Pot: A small bonsai container with drainage holes is included. It is essential to house the bonsai plant and providing the right conditions for growth.

3. Bonsai Soil Mix This kit contains a ready-mixed bonsai soil specifically designed to ensure proper drainage and aeration. Both are crucial to the health of bonsai trees.

4. Bonsai Wire: It's possible to find bonsai wire of various sizes to create and shape the branches of your tree. Wiring is one of the most fundamental techniques for bonsai styling.

5. Pruning Shears: Two pairs bonsai pruning tools is provided for shaping and trimming the tree's branches as well as its the foliage.

6. Bonsai Instructions: In-depth directions or a book for bonsai styling and care methods are usually included to assist you in getting started.

7. Tips on Training and Styling The guidebook may offer tips for how you can dress and instruct your bonsai tree with guidance regarding the proper placement of branches, pruning as well as wiring.

8. Fertilizer: Certain kits contain some bonsai-specific fertilizer that will help to nourish the tree in a proper way.

9. Bonsai tools: Some kits could include equipment such as a root hook as well as scissors or wire cutters. These are useful for higher-end methods.

10. Small Watering Can: A container or nozzle could be part of the package to help properly water your bonsai.

11. Bonsai Display Stands or Trays There are kits that include the option of a tray or stand that can showcase your bonsai. This can add the aesthetics to the setup.

12. Decorate with rocks or pebbles for covering the surface of the soil, and to enhance the appearance to your bonsai.

If you are using the DIY Bonsai Kit, it's crucial to follow the directions carefully and pay careful attention to the particular requirements of the bonsai trees contained within the kit. Remember that cultivating bonsai is a continual training process. As your knowledge improves and you improve, it's possible that you'll want to investigate more advanced methods and other tools to improve the appearance of your bonsai trees.

Chapter 7: The Basics Techniques

Pruning and wiring are the most important practices in the cultivation of bonsai. They play an important role in shaping and sustaining the appearance of bonsai trees. This article will provide an overview of the techniques used:

Pruning:

1. The purpose of pruning is to specific removal of branches roots, or leaves in order to create the ideal dimension, shape, and design of bonsai trees. Pruning is a way to ensure the balance and health of the bonsai tree.

2. Types of Pruning:

Maintenance Pruning: Regularly eliminate dead or dying leaves, growth that is excessive, as well as branches that are not needed to maintain the form of the tree.

O Hard Pruning Stiff cutting of foliage or branches to stimulate back budding and to create new designs.

Pinching: Carefully taking out terminal buds in order to promote the growth of branches and thicker leaves.

3. Instruments: Pruning shears, or cutters are utilized to create precise cuts, without damaging the trees. Concave cutters can be utilized on larger branches.

4. Techniques:

Prune above an apex or leaf to encourage an increase in growth.

Use angles for more aesthetics and healing.

Maintain the equilibrium for the trees in your mind when pruning.

5. Timing: Pruning is usually carried out during the growing seasons generally in spring or in the early summer months during the time when trees are developing. However, it is possible for maintenance pruning to take place all the time.

Wiring:

1. It is the art of twisting or shaping trees and branches to attain the bonsai design. Wiring allows you to achieve motion, balance and harmony within the structure of the tree.

2. Types of Wire:

O Aluminum Wire: It's soft and simple to handle. It is typically found on evergreen trees.

Copper Wire: More durable and holds branches better. It is used on conifers as well as larger branches.

3. Techniques:

The wire should be wrapped over branches in an angle. Make sure there is sufficient space between the coils to allow the branch to expand.

Begin at the bottom before moving towards the top.

O Bend branches gradually and slowly to avoid sharp angles that may damage bark.

The wire should be taken off when it is set to put the branch at the proper position. This usually happens within a couple of months.

4. Wiring Guidelines:

Wire is used during the winter months for deciduous trees, and in the season of growth for conifers.

Be careful with smaller and more delicate branches in order to prevent the risk of injury.

Make sure to check and alter the wires regularly in order to avoid scratching as the branch gets thicker.

5. Security: Take care when you wire to prevent damaging the trees blood vessels. Make sure to protect the bark with Raffia tubing or rubber under the wire.

6. Styling: Wiring enables the creation of classic bonsai designs such as traditional uprights, casual uprights cascades and much more.

The two processes need practice, as well as an attention to aesthetics. When you've gained experience becoming more skilled in using these methods to form and shape your bonsai plants, making the trees into living artworks.

The steps to take to prune and wire

Pruning and wires are crucial practices in the care of bonsai and style. These are the steps to follow to follow for both

Pruning:

1. Gather Your Tools:

Make use of bonsai pruning shears and scissors to make precision cutting.

Check that your equipment is cleaned and disinfected to stop the spread of infectious diseases.

2. Identify What to Prune:

Analyze your bonsai plant to identify what branches, leaves or branches require trimming.

Concentrate on getting rid of the dead, diseased, or yellowing leaves first.

Check for branches that are threatening the desired form or equilibrium.

3. Plan Your Cuts:

Determine where you will cut. To prune branches you should cut only above the branch or a node.

Keep in mind the overall style and design you'd like to attain.

4. Make Clean Cuts:

You can make a one-time fast cut in order to prevent breaking or damaging the tree.

Cut just a bit away from the leaf or bud for better healing and to prevent the spread of.

For branches with more thickness use concave cutters make an equilateral cut.

5. Dispose of Pruned Material:

Take off the branches that have been pruned and foliage out of the bonsai pot in order to avoid rot and insects.

Dispose of, or compost the cut material.

6. Monitor and Adjust:

Check your bonsai regularly to see if it has new growth, and modify your pruning when needed in order to preserve the desired shape and equilibrium.

Be careful not to over-prune because it could cause the tree to weaken.

Wiring:

1. Select the Right Wire:

Select the right gauge and wire (aluminum or copper) for the bonsai you want to grow as well as the branches you wish to form.

The wire should measure approximately one-third of the width of the branch you plan to bend.

2. Prepare the Branch:

Check the branch you're planning to connect, making sure the branch is healthy and free of insects or disease.

If the branch appears too rigid or thick you may want to consider making use of guy wires, or gradually stretching techniques prior to the principal wire.

3. Anchor the Wire:

Begin at the top of the branch, and then carefully tie the wire around it by slanting it slightly.

There should be sufficient space between coils of wire for the branch to expand.

4. Shape the Branch:

Gradually, and slowly move the branch until it is in the desired angle.

Be patient, and beware of abrupt, sharp bends that can damage barks or the vascular tissues.

5. Secure the Wire:

Wrap the wire over the branch's new location and secure it by anchoring it.

o Make sure the wire is secure enough to hold the branch, but not too tightly securing it to prevent harm.

6. Check and Adjust:

Inspect regularly branches that are wired to stop wiring from cutting into bark of the tree as it grows.

Take the wire off after it has placed the branch to the position you want, typically within a couple of months.

7. Protect the Bark:

To stop wire from scarring by rubbing, put raffia or rubber tubing under the wire when it comes into contact with the branch.

8. Plan and Execute:

A Wire that has a clear strategy in your mind, taking into consideration the overall look and feel that you want your bonsai to have.

Pruning and wiring both need practice and experience, so it's important to pay attention to your tree's aesthetics and health when you're performing these tasks. As time passes and you gain experience it will become easier with shaping and sustaining your bonsai trees.

Chapter 8: Designing Your Bonsai

How to style your bonsai

How to style your bonsai is an artful and imaginative method that involves creating and refining the shape of the tree in order to create a pleasing appearance. Below are some steps for how to decorate your bonsai

1. Understand Bonsai Styles:

Learn about different bonsai designs, including traditional uprights, casual upright semi-cascade, cascade windspread, literati and windswept. Each has its own distinct characteristics and the design principles.

2. Determine the Style:

Choose the style you'd like to create by analyzing the bonsai's features like the type of tree, the form of the trunk, as well as branch form. Your bonsai's natural characteristics are the best guide for your decision.

3. Evaluate the Tree:

Examine your bonsai's current health, taking into account the size of its trunk, its branch position, as well as general condition. Find features that are in line with the design you have chosen.

4. Select the Front:

Pick the front view angle that emphasizes trees best attributes and compliments the chosen style. The front is the most appealing visual view.

5. Pruning and Branch Selection:

Trim away any unwanted branches, leaves and new growth in order to shape the trees. Eliminate branches that do not conform to the chosen style, or impede the overall appearance.

Make sure to keep branches that contribute to your desired shape and harmony.

6. Wiring:

Make use of bonsai wire to shape the trunk, branches, and leaves to fit the style you have chosen. Wire should be positioned to create natural looking curves and motion.

Keep your patience and stay clear of excessive bending as it could cause damage to the tree. Be sure to adjust and check the wire on a regular basis.

7. Branch Placement:

Be aware of the arrangement of branches relationship to the trunk. They must flow in a harmonious manner and form a harmonious silhouette.

Ensure that branches branch from the trunk in appropriate lengths and angles.

8. Apex and Taper:

Determine the Apex (the highest point) on the tree so that it will match the chosen style. The apex must be clearly defined and play a role in the overall balance of the tree.

Make a taper, ensuring that branches and the trunk diminish in their thickness gradually when they are removed of the base.

9. Patience and Iteration:

It's a continual procedure. Be patient and don't overdo it. The tree will adjust to changes over the course of time.

Be sure to regularly check the growth of your tree and make modifications as required to preserve the design.

10. The Balance and Harmony Try to achieve balance and harmony within your bonsai's style. Do not overcrowd branches and also ensure that your components of the tree work in harmony.

11. Finess and detail: Concentrate on more finer points, like refining the leaves pads as well as creating natural Jin (deadwood) and also increasing the necari (surface root). Make use of specific tools and techniques to achieve sophisticated styling techniques, like the carving process, defoliation, and deadwood works.

12. Constant Maintenance: keep the shape of your bonsai by trimming and wiring it when it

expands. Keep track of the tree's health and deal with any issues quickly so that it can maintain its beauty.

A bonsai's style is an ongoing and dynamic method that changes over the course of time. It requires an sense of aesthetics, an knowledge of the design's fundamentals, as well as an appreciation of the practice of bonsai. When you've gained experience becoming more skilled in shaping and refining the bonsai tree to make life-like artworks.

Pruning methods

Pruning Techniques in Bonsai and Promoting Rampant Growth:

Pruning Techniques:

1. Maintenance Pruning:

It is important to regularly remove dead, dying, or ill-healthed branches and leaves to ensure the bonsai's beauty and health.

Restrict growth in order to promote the process of ramification (the creation of smaller branches).

2. Hard Pruning:

O Hard pruning is the greater drastic cutting of leaves or branches in order to shape the bonsai.

o It encourages back budding that encourages the growth of new branches and shoots close to the tree.

3. Pinching and Nipping:

Utilize your fingers or bonsai scissors to pinch or cut the buds at the end of branches for a more dense and branching leaves.

4. Candle Pruning (For Conifers):

o On bonsai with coniferous species like trees like spruces or pines, choose to take off the long-elongating branches (candles) throughout the season of spring in order to manage the growth of the tree and its shape.

5. Directional Pruning:

Prune the branches in the direction you wish they to develop. This helps in guiding the development of your tree as per your preferred layout.

6. Leaf Pruning (For Deciduous Trees):

To decrease the size of leaves and promote ramification, you can selectively reduce or eliminate certain leaves in the growth season.

Promoting Rampant Growth:

1. Fertilization:

Provide a balanced fertilizer for bonsai in the growth season (spring through the summer) to make sure your tree is getting the essential nutrients to ensure a vigorous development.

2. Watering:

Keep consistent and proper methods of watering to maintain a uniformly moist soil. Proper hydration is vital to healthy growth.

3. Sunlight:

Ensure that your bonsai is receiving enough light based on the species it is. The majority of bonsai thrive under direct, bright lighting. The right amount of sunshine is crucial to photosynthesis as well as a robust development.

4. Air Circulation:

Air circulation is essential to prevent stagnation and supports the growth of healthy plants. Beware of overcrowding the foliage or branches.

5. Regular Pruning and Training:

Always trim and instruct your bonsai's growth to be directed to where you would like to direct it. Regular maintenance helps prevent growth, which can be sloppy and insurmountable.

6. Repotting:

Repotting each 2-3 times (depending on the plant species) helps to refresh the soil and stimulates the growth of roots and improves overall well-being and strength.

7. Pest and Disease Management:

It is imperative to address any insect problems or illnesses that could affect the growth. A healthy bonsai grows more quickly.

8. Stress Reduction:

Reduce stress-related factors such as extreme temperature variations, excessive watering and underwatering. Stress can impede growth.

9. Balanced Styling:

o Make sure the design of your tree is in balance and permits an appropriate growth distribution among leaves and branches.

10. Prune for Direction:

Prune branches to direct development in the right direction for aesthetic reasons or to help balance the tree's design.

Be aware that encouraging rapid growth must be in line with your bonsai's overall style and design ambitions. The growth rate is important, but it must be managed to preserve the aesthetics and proportions of your bonsai plant.

Advanced Bonsai Styling

Modern bonsai styling techniques extend beyond the basic methods to make intricate and captivating bonsai trees. Here are some of the most advanced techniques for styling:

1. Defoliation:

The process of defoliation is the removal of most or all leaves in the springtime.

o It may encourage back budding, shrink leaf size and provide an elegant appearance.

It is commonly used for evergreen and deciduous trees.

2. Deadwood Techniques:

Deadwood techniques like the jin (stripped bark) and shari (carved deadwood) to create an appearance of aging and character.

o They can be utilized to simulate the natural aging and weathering processes on bonsai branches and trunks.

3. Surface Roots (NIBARI):

The development of surface roots, or NIBARI increases the bonsai's strength as well as its aesthetic appeal.

Expose and strengthen the surface root system by making sure you are careful with your pruning, root placement and soil control.

4. Bonsai Trunk:

Modern styling of bonsai trunks is the creation of taper, movement as well as textures.

o Methods like carving, using power tools, or approach Grafting are a few ways to improve the shape of the shape of the trunk.

5. Bonsai Forest:

The bonsai garden is a collection of trees that are planted in a group to resemble a natural forest.

Advanced styling entails balance of the heights of trees and arranging them into an harmony, and resembling the pattern of growth that is typical of the tree.

6. Bonsai Rock Planting:

A bonsai rock garden incorporates rocks into the bonsai design.

The most advanced techniques include selecting and then arranging stones in the natural look of a landscape while also incorporating bonsai trees in a seamless manner.

7. Advanced Wiring Techniques:

Advanced wiring entails the creation of complex detailed branch and foliage configurations.

Methods such as guy wiring (using other support wires) and multi-wire wiring as well as fusion methods can be used.

8. Grafting Trees:

It is also utilized to grow more branches or variety of bonsai plants.

Advanced techniques for grafting like thread grafting as well as approach grafting allow exact control of the design of the tree.

The advanced techniques need a thorough comprehension of bonsai horticulture as well as art. You must approach these techniques carefully and test them with less expensive trees prior to trying them on the most prized of specimens. In addition, sophisticated styling must be in line with the nature of the tree as well as the style of bonsai you'd like to see.

Chapter 9: Different Styles Of Bonsai

Bonsai provides a variety of sub-styles and styles, each one with their own distinct characteristic and style. Below are a few of the most well-known bonsai designs and sub-styles of them:

1. Formal Upright Style (Chokkan):

Classical Chokkan Straight and upright trunk, with slowly decreasing branches' thicknesses.

Moyogi: Lightly curving trunk, while still maintaining an upright look overall.

2. Informal Upright Style (Moyogi):

Classical Moyogi: Elegant and slightly curled trunk, with well-balanced branches.

Windswept (Fukinagashi) Windswept (Fukinagashi): The tree seems to be formed by the force of winds. The branches that lean in a particular direction.

3. Slanting Style (Shakan):

Literati (Bunjin) Literati (Bunjin) artful and expressive style featuring a an irregular, contorted trunk.

Cascade (Kengai) The tree's the trunk cascades down, usually mimicking a tree that is growing on the cliff.

4. Semi-Cascade Style (Han-Kengai):

Semi-Cascade: a compromise between cascade and upright, with a trunk that partially cascades.

5. Broom Style (Hokidachi):

Official Upright Broom (Chokkan Hokidachi) Straight like a broom, it has well-spaced branches.

Informal upright Broom (Moyogi Hokidachi) More relaxed and natural style, but still maintaining the form of the broom.

6. Literati Style (Bunjin):

Windswept Literati (Fukinagashi Bunjin) is a mix of windswept and literati designs, usually featuring a bent and twisting trunk.

7. Multi-Trunk Style (Ikadabuki):

Forest (Yose-ue) The term "forest" refers to a group of trees that are planted to create the appearance of an actual forest.

Clump style (Ikadabuki) The concept of multiple branches that emerge from the same root system, forming an extremely large cluster.

8. Group Planting (Yose-ue):

Multi-Tree Planting (Ikadabuki) The trees are with different varieties or species placed together.

Multi-Level (Ikadabuki) trees placed on different levels in order to give depth and attraction.

9. Literati Cascade (Bunjin Kengai):

A combination of the styles of cascade and literati with the cascading trunk and contorted.

10. Windswept Cascade (Fukinagashi Kengai) It is a blend of styles like cascades and windswept that creates the look of a tree formed by the force of strong winds.

11. It is a Bonsai-inspired rock plantation (Ishitsuki) (Ishitsuki): The Waterfall (Taki Zumi): Trees are planted on rocks in order to mimic the look of a waterfall. Cliff (Ganban Zumi) trees are incorporated with rocks in order to create a rock-like scene.

12. Multi-Style (Ikkan) Multi-Style (Ikkan): a single tree with traits of different styles in the same tree.

Bonsai styles, as well as sub-styles offer endless possibilities for imagination and expression. If you are styling your bonsai take into consideration the natural features of your tree and select the style that highlights the aesthetics of the tree, while also reflecting your own personal style as an artist of the bonsai.

TIPS AND TRICKS

The best bonsai outcomes require an understanding of horticulture along with artistic skills and perseverance. Below are the most important and most effective tips and techniques for the success of bonsai gardening:

1. Choose the Right Tree:

Choose a tree species which is suitable for the climate you live in and also your degree of experience as a novice.

It is best to start with a plant that displays desirable bonsai traits for example, tiny leaves as well as a strong branch form.

2. Quality Soil Mix:

Make sure to use a well-drained bonsai soil mix for an adequate amount of aeration as well as water retention. Do not use garden soil as it could cause inadequate drainage.

3. Proper Pot Selection:

Pick a bonsai planter to match the style and dimension. The pot must have drainage holes as well as fit the correct size for the root system of the tree.

4. Understand Watering Needs:

Make sure to water your bonsai well and allow the soil time to dry completely between watering to stop the root from rotting.

You can adjust the frequency at which you water according to the weather conditions and your trees' specific needs.

5. Adequate Sunlight:

Provide the appropriate amount of sunshine depending on the type of tree you have. Bonsai grow best in light that is bright and indirect.

The tree should be rotated regularly to guarantee even growth and to prevent development that is one-sided.

6. Regular Pruning and Training:

Train and hone your bonsai frequently to keep its form and promote the process of ramification (branch growth).

Make use of wiring techniques to direct the growth of the tree and to create the desired shape.

7. Fertilization:

Apply a balanced and balanced bonsai fertilizer throughout the growing season in order to ensure that you get the nutrients necessary to ensure healthy growth.

Adjust the frequency of fertilization depending on the type of tree and the specific needs.

8. Patience and Observation:

Bonsai is a form of perseverance. Be attentive to your tree and take decisions in based on the response of your tree to pruning, and the instruction.

Do not rush the growth process as it requires the time to develop and improve its design.

9. Repotting:

Repot your bonsai each two to three years (varies according to the species) to fertilize the soil, cut roots, and encourage strong development.

Repotting can also be an opportunity to alter the position of the tree within the pot, and to improve the condition of nebari (surface root).

10. Management of pests and Disease Management: - Check your bonsai regularly for any signs of pests or disease. Make any necessary adjustments in order to stop destruction. In the event of an infection, isolate affected trees to stop the spread of disease to bonsai in other areas.

11. Guard against extreme conditions Protect during winter: Provide winter-time protection for bonsai that are cold sensitive in cold conditions. The bonsai should be shaded and

protected when temperatures are extreme to protect them from burns and loss of water.

12. Develop and adapt: Always keep yourself informed about how to care for your bonsai as well as methods. Join workshops, attend clubs for bonsai, and also read books and other articles. You can adapt your strategy according to the particular requirements and traits of your bonsai plants.

13. Maintain Balance: Bonsai requires balance. Create a harmonious connection between the tree's trunk branches and the foliage. Don't overload or crowd branches because this could harm the health of your tree as well as its appearance.

14. Have Fun on the Journey Bonsai is a lifetime experience. Enjoy the process, take lessons from your failures and successes and be proud of the ever-changing beauty of your bonsai designs.

15. Seasonal Adjustments: Be aware that the care of bonsai plants varies with the seasons.

Make adjustments to your fertilization, watering and routines for protection depending on the time of the year.

16. Soil Testing: Periodically check the soil of your bonsai to make sure it has the desirable qualities. The easiest test to do is a percolation test in order to verify the drainage.

17. Aesthetic Pruning: When pruning, think about the appearance that the trees have. Try to achieve a balanced branching distribution attractive foliage pad layouts as well as natural-looking forms.

18. Dressing the soil surface: Make use of decorative moss or tiny pebbles for covering the soil's surface and improve the overall look that your bonsai will have.

19. Air Layering: Explore methods of air layering to spread and grow new bonsai plants out of existing trees.

20. Seasonal Color Display Pick bonsai species with seasons-specific color variations like fall leaves or winter blooms for a visual appeal.

21. Pest Control: Use methods to prevent pests, such as Neem oil or insecticidal soaps to stop insects that are common to your bonsai.

22. Learn from masters: Learn from the work of famous bonsai masters. Try to imitate their methods and designs.

23. Be patient when wiring: If you are wiring it, be patient to make perfect curvatures and bends. Do not rush through the process as well as be patient when working with the branches of your tree.

24. Think about the color of your pot: Select bonsai containers that match the color and design of the tree. The earthy hues are often a good fit However, contrast can make a statement.

25. Bonsai Photography: Document the growth of your bonsai using regular pictures.

This helps you assess your methods and observe the growth of the tree.

26. Mindful watering: - The bonsai's water source is from above to mimic natural rain. This method helps to wash away dirt and insects and helps in hydrating the soil.

27. Adjusting the pH of your soil: Check the soil's pH and alter it as required in order to satisfy the requirements for your particular tree species.

28. Bonsai grooming tools: Make sure you invest in grooming tools of high-end quality including concave cutters and knob cutters, as well as sharp-tipped scissors to ensure precision.

29. Participate in Bonsai Exhibitions: Attend bonsai exhibits to get inspired and see stunning specimens and meet other fans.

30. Teaching and Sharing: The act of teaching others about bonsai will increase your knowledge and understanding of this

practice. You might consider teaching your bonsai knowledge to fellow fans.

31. Be open to imperfections. - Keep in mind that perfection isn't the aim of bonsai. Enjoy the distinct nature and imperfection of your trees as they add to the beauty of your garden.

32. Personal Connection: Develop an emotional connection to your bonsai. Give them your love and love and they'll react positively.

33. Explore and innovate: Be open to experimenting with different techniques, styles and material. Innovative thinking can produce stunning and unique bonsai designs.

Keep in mind that bonsai isn't only about making tiny trees, it's also an art of nurturing and sustaining living things. If you are committed and persistent it is possible to achieve outstanding outcomes, and build a strong admiration for bonsai's art.

Below are the answers to a few frequent questions and worries novices might have regarding bonsai

1. What is the best frequency to be able to water my bonsai?

The frequency at which you water will depend on variables such as the species of tree, size of pot, the weather, and the season. The general rule is to you should water your bonsai at the time that the soil's top layer is a little dry. Don't let it dry entirely or get waterlogged.

2. What kind of soil do I need to use to grow my bonsai?

Choose well-drained bonsai mixes that offer good air circulation as well as conserving water. Do not use regular garden soil as it may cause low drainage and root-related issues.

3. How can I select the best bonsai container?

Choose a container that is compatible with the tree's design and dimensions. It should have drainage holes, and it is appropriate to the roots. Bonsai pots come in many forms and materials. So select one that matches the aesthetics of your tree.

4. How can I keep my bonsai inside all year round?

Some bonsai species, such as Ficus and Jade are able to thrive inside However, the majority of bonsai thrive being outdoors in order to get natural sunlight as well as seasonal temperatures. Take into consideration the particular requirements of the species you have chosen for your tree.

5. Does pruning need to be done for bonsai care?

Pruning is an essential part of the care of a bonsai. Regularly pruning maintains the shape of the tree, promotes the process of ramification (branch growth) while keeping the bonsai healthy.

6. What can I do to protect my bonsai against pests and illnesses?

Check your bonsai frequently to identify signs of a pest or disease. Make use of preventive measures such as Neem oil or insecticide soap to repel insects. If problems do arise, address immediately to avoid destruction.

7. How often should I pot my bonsai?

Bonsai generally require repotting every 2 to 3 years in order to replenish the soil, cut back the roots and encourage the growth. The time to repot varies based on the species and size of pot.

8. Do I have the ability to create the bonsai of any type of tree?

If you want to try to make bonsai from a variety of tree species, you should start with ones that are known for their versatility for beginners, like Ficus, Juniper, or Chinese Elm. Certain species are best suited for those who are new to bonsai because of their tolerant nature.

9. What can I do to make my bonsai look more attractive? an ideal shape or style?

Style involves pruning and wiring in order to create a unique bonsai. Learn about the look you'd like to achieve and then follow the appropriate techniques. The patience and the gradual adjustment are crucial.

10. What causes bonsai trees to have branches or leaves fall off? Loss of branches and leaves could be due to a variety of reasons, such as the environment and root-related issues, as well as bugs, or poor treatment. Find out the root cause of the problem and then take appropriate steps to correct the issue.

11. Can I display my bonsai indoors during winter? It's typically recommended to display the majority of bonsai outside. You can however offer protection against freezing temperatures as well as strong wind to protect your bonsai from injury.

12. What can I do to thicken the tree's trunk? tree? The process of thickening the trunk is an arduous process, and is most common in the first phases of development. The uncontrolled growth of a tree and the slow pruning will help in thickening the tree's trunk as time passes.

13. Do I have the ability to grow bonsai using cuttings or seeds? You can make bonsai with cuttings or seeds, however it's a long process than starting from already-existing bonsai plants. Seeds allow the tree to be completely in control over your tree's growth.

14. Do I have to set a date for wiring my bonsai? The wiring process is usually done during the growth season during spring and summer in the spring and early summer, when branches are malleable. Be sure to avoid wiring in the winter months so that the tree is not damaged.

15. What can I do to protect my bonsai against extreme climate conditions? Make use of measures to protect it, such as shade

cloths during extreme temperatures and winter-time protection with mulching or wrapping during freezing temperatures to protect your bonsai.

16. Do I need to use tap water for watering my bonsai? Tap water is generally suitable for bonsai. However, you should let it remain for at least one couple of days for chlorine to evaporate. For areas with hard water, the occasional application of distilled or rainwater will help to prevent the build-up of minerals within the soil.

17. What do I do when my bonsai is in need of to be repotted? Examine the bonsai's roots in the time of repotting. If your roots are thickly packed and circling around the pot or your soil doesn't hold water effectively, then it's the time to repot it.

18. What is the distinction between bonsai care for outdoor and indoor maintenance? The indoor bonsai need more regular temperatures and humidity levels. Outdoor bonsai can be exposed to nature's elements

and could require protection from extremely cold conditions. They both require adequate light and watering, however the exact maintenance is different for each species.

19. How can I make a bonsai tree that I saw in the wilderness or in my backyard?

Yes, you can take wild or even gardens trees to make bonsai, nevertheless, you must adhere to the ethical guidelines and laws. Furthermore, not all tree species are appropriate to be used for bonsai. This is why it is important to conduct research.

20. Does bonsai constitute a type of cruelty to plants?

When done in a proper manner the bonsai tree isn't harmful for trees. Bonsai trees are meticulously cared for and maintained in order to maintain their longevity and health. This is the art of making an image of the growth and development through the nature.

21. How can I ensure the equilibrium between roots and leaf growth?

Regular trimming and pruning of the roots when repotting helps maintain the balance. Do not allow the leaves to expand too quickly when compared with the roots.

22. What can I do if my bonsai tree begins to lose its form?

If your bonsai has lost its shape because of overgrowth or poor pruning, you are able to gradually restore it's form with meticulous pruning and wiring. Find guidance from knowledgeable bonsai enthusiasts, if you require.

23. Are there ways to plant multiple bonsai plants in the same pot?

It is possible to build the appearance of a forest with multiple trees planted in the same area. Make sure that the trees that you select are suitable both in terms of their growth rate and aesthetics.

24. What can I do to encourage blooming on my flowering bonsai?

Create the ideal climate conditions, which include the right temperature and lighting specifically for your species. Some bonsai that bloom require an extended period of inactivity before they begin blooming.

25. Do I need to fertilize my bonsai all year long?

You can fertilize your tree all year long however, bonsai generally require less or no fertilization in the time of dormancy, usually the winter months. Change the timing of fertilization depending on the specific requirements of your tree.

26. What's the purpose of moss in the bonsai culture?

Moss is a great addition to the look of your bonsai as well as aid in retaining the moisture in your soil. Additionally, it provides an organic, natural look on the ground on the trees.

27. What can I do to prevent my bonsai from growing unbalanced (long branches that have sparse leaves)?

Pruning regularly and squeezing new growth back is a good way to prevent lagging. Make sure to create a dense healthy, well-balanced canopy.

28. What is the best way to determine the date of the bonsai's birth?

Bonsai age determination is not easy due to the fact that it is influenced by factors like the growth environment and methods applied. Bonsai trees are admired more because of their beauty and charm as opposed to their actual old age.

29. What can I do if the leaves of my bonsai change color, turning brown or yellow?

Browning or yellowing leaves may be a sign of a variety of issues, such as the over-watering of your garden, the underwatering of your property and pests as well as ailments. Examine the causes and then take the appropriate steps, for example, adjusting watering levels or dealing with pest issues.

30. Do I have the possibility of propagating my bonsai through cuttings?

There are bonsai varieties that are able to be propagated using cuttings. It is a great option to make new bonsai plants or make a replacement for an older bonsai by an older specimen.

Keep in mind that bonsai is an art which requires practice and learning. Be resilient and don't let setbacks discourage you Continue to learn about the particular requirements for your tree. joining a bonsai group or seeking guidance from seasoned bonsai experts can prove beneficial when you begin the bonsai path.

Chapter 10: Principal Resources

Below is a listing of reputable sites as well as blogs on which you can obtain additional information, tips as well as resources on bonsai.

1. Bonsai Empire (https://www.bonsaiempire.com/)

An extensive resource that includes tutorials, articles as well as the online bonsai training course.

2. Bonsai Nut (https://www.bonsainut.com/)

A bonsai community that is popular that lets enthusiasts talk about their experiences and get assistance.

3. Bonsai Tonight (https://bonsaitonight.com/)

A blog written from the bonsai master Jonas Dupuich, featuring articles as well as videos and instructional material.

4. Bonsai Empire YouTube Channel (https://www.youtube.com/user/bonsaiempire)

The site offers video tutorials, demonstrations and talks with bonsai experts.

5. Bonsai Resource Center (https://bonsairesourcecenter.com/)

Provides guides, articles as well as product reviews that relate to the care and maintenance of bonsai trees.

6. BonsaiMary (https://bonsaimary.com/)

Provides information and articles regarding bonsai tree care and cultivating.

7. International Bonsai Magazine (https://www.internationalbonsai.com/)

Online magazine that includes gallery galleries, articles and other events that are associated with bonsai.

8. American Bonsai Society (https://absbonsai.org/)

Official website for the American Bonsai Society, which offers educational materials and the publication of a publication.

9. Bonsai Empire Forum (https://forum.bonsaiempire.com/)

A bonsai forum in which fans can post questions, exchange experiences and gain knowledge from one another.

10. Bonsai Tonight YouTube Channel (https://www.youtube.com/user/BonsaiTonight)

Videos on different aspects of care for bonsai stylistic, styles and methods.

11. Bonsai Empire Blog (https://www.bonsaiempire.com/blog)

It contains posts on bonsai related topics.

12. Bonsai Society of Greater New York (http://www.bsgny.org/)

Information on upcoming the latest events, workshops and bonsai-related resources throughout New York. New York area.

13. The Bonsai Journal (https://thebonsaijournal.com/)

The journal is online and focused on the art of bonsai with galleries, articles as well as other information.

14. Bonsai Tree Forums (https://www.bonsaitreeforums.com/)

A forum that is a community driven platform that allows bonsai lovers to talk about different topics and ask for suggestions.

15. Bonsai Club International (https://www.bonsai-bci.com/)

It also provides a publication, details, and resources about bonsai groups around the world.

16. Bonsai Empire Facebook Group (https://www.facebook.com/groups/bonsaiempire)

A group on Facebook in which bonsai lovers can meet and share their experience as well as ask questions.

Additional resources are listed below:

1. Bonsai Organizations and Clubs If you are interested in joining a local bonsai group or club could be an excellent source for learning hands-on through workshops and networking to experienced bonsai lovers. Find clubs in the area you live in or around to join.

2. Bonsai Books: There are several bonsai-related books available which provide detailed information such as specific species-specific guides to care stylistic techniques as well as artistic insight. The most highly-respected titles are "Bonsai Basics" by Colin Lewis, "The Complete Book of Bonsai" by Harry Tomlinson, and "Bonsai Techniques I & II" written by John Yoshio Naka.

3. Bonsai workshops and classes There are many botanic gardens, nurseries and bonsai experts provide classes and workshops.

Engaging in these workshops will provide invaluable experiences and guidance that is tailored to your needs.

4. Bonsai exhibitions and shows: Participate in bonsai shows and exhibitions within your local area, or on the national level if it is possible. They showcase outstanding bonsai trees as well as a chance to learn the methods and styles employed by expert bonsai artists.

5. online Bonsai Communities: Browse the internet for forums and social media communities that are dedicated to bonsai. Sites such as BonsaiNut, The BonsaiChat website, as well as Reddit's Bonsai subreddit provide forums to ask questions, share stories, and getting tips.

6. Bonsai Magazines: Sign up to bonsai publications like "Bonsai Focus," "Bonsai Today," and "International Bonsai." The magazines feature stories on bonsai artists' profiles, interviews and news about the latest developments in bonsai.

7. Bonsai Nursery Visits: Visit local bonsai nursery to look at the variety of bonsai plants in close proximity and talk to experts who are able to give advice and guidance.

8. Online Bonsai Classes: Many websites offer bonsai classes and can be an ideal way to study at your own speed. Sites such as Udemy and Skillshare frequently offer bonsai-related classes.

9. Bonsai Videos as well as YouTube channels: YouTube has a variety of bonsai related channels with experts sharing tutorials, videos as well as knowledge. The most popular channels are "Bonsai Mirai" by Ryan Neil as well as "Bonsai Empire" for video tutorials.

10. Bonsai Software and Apps: Think about using bonsai design or management tools or apps for mobile devices which aid in scheduling, monitoring care routines and the styling of your bonsai. They can be useful to those who enjoy the digital tools available to them.

11. Bonsai auctions and sales: Look into bonsai auctions, sales and activities, which could provide an chance to buy rare specimens, pots or any other bonsai related items. Make sure you study and comprehend the qualities and worth of the items prior to engaging in the event.

12. Bonsai Podcasts There are a variety of podcasts devoted to bonsai. They feature interviews with experts, debates regarding different aspects of bonsai and tips to those interested in the art. For instance "Bonsai Wire" and "Bonsai Network Podcast."

13. Local Bonsai Gardens There are some cities that have open bonsai parks or arboretums which showcase stunning bonsai collections. Visits to these gardens could inspire and provide insight on the cultivation of bonsai.

14. Bonsai Photography: Read the books and other resources for bonsai photography, and discover how to record the beautiful beauty of your bonsai plants with breathtaking

photos. Photos of high-quality are useful in documenting progress and showing your work.

15. Botanical and Horticultural Libraries Local libraries, along with the botanical and horticultural institutes might have large collection of journals, books as well as other materials associated with bonsai.

Utilizing these extra sources, you will be able to expand your understanding, broaden your bonsai knowledge base, connect with other members of the community, and improve your knowledge and enthusiasm of this engaging art.

Note that although the blogs and websites listed can be valuable sources of details, it's important to check cross-references and refer to several sources in order to have a complete and comprehensive knowledge in the area of bonsai.

Chapter 11: The Bonsai Art

The art of living called bonsai can be described as beyond just gardening. It captivates the heart as well as the mind.

We embarked in a search to understand the essence of bonsai by exploring its tangled web of significance to the culture theology, philosophy, and historical context.

What exactly is bonsai?

In Japanese it is believed that the term "bonsai" means "planted in a container." The simple explanation however is not enough to convey the intricate craft and deep significance Bonsai represents justice. The premise of bonsai is the art to grow tiny trees within pots that are meticulously shaped and pruning them so that they resemble the larger trees in nature.

In this part, we'll look at the basic aspects of bonsai in this segment. Learn about the art of miniatureization, and learn how bonsai experts are able to capture the grandeur of trees that are tall inside tiny containers.

Also, we will discuss the importance of culture in bonsai and its significance in ensuring peace between people and nature.

It is impossible to define bonsai by its own It is a declaration of dexterity, rigor as well as unwavering reverence to the beauty and vitality that is inherent in all trees.

1.2 The Origins of Bonsai and Its Philosophy

An exciting journey through the ages as well as across different civilizations can be discovered in the rich background of bonsai. While bonsai is often associated with Japan but its roots could be traced back to early China which is where the technique in "penjing," or the cultivation of tiny landscapes as well as trees, set the stage for the creation of bonsai.

It was however in Japan that bonsai began to become an art form that is internationally recognized.

We will explore the historic growth of bonsai during this portion. We will follow the evolution of bonsai through the time of prehistoric China through modern-day Japan in which it was affected by both Zen Buddhism and the Bushido code of samurai.

The bonsai art form evolved from a mark of status and wealth to becoming a kind of art that is accessible to people with all kinds of different backgrounds.

Then we'll examine the philosophical concepts which underlie bonsai like wabi-sabi. It is a celebration of the beauty of the process of change and imperfections.

The deeper meaning the art form has can be achieved by knowing the history and philosophy behind bonsai. Then, we will gain an knowledge of the significance of bonsai as well as its lessons about our connection to nature throughout this section.

The Evolution of Bonsai: Its Philosophy

Explore the intricate network of influencers that created bonsai throughout the many centuries, while we continue to look into its origins and the philosophy behind it.

Bonsai and Zen Buddhism: A Spiritual Connection

Zen Buddhism has had a substantial influence on bonsai. Zen practitioners emphasized on the pursuit of awareness and the benefit of what is happening in the moment in search of enlightenment as well as calm in the most fundamental objects. By focusing on balance, harmony along with the process of nurturing bonsai is very similar to Zen notions.

Zen monks as well as academics helped make bonsai a popular form of meditation. They saw taking care of Bonsai trees as an act of meditation. It was a way to connect with nature and also a means to achieve an inner calm.

Zen Buddhist monasteries' bonsai plants evolved into centres of contemplation and illumination with the exact form and trim of

The trees were used as a symbol to show the discipline and focus required to progress spiritually.

A Code of Ethics for Bonsai and Bushido

Additionally to Zen Buddhism, Bushido, the code of warriors is a major influence on bonsai. Bushido is the method of a warrior, places the importance on virtues such as respect, loyalty, honor and control over oneself.

The arduous maintenance and care of trees was thought to be the mirror to the ideals of bonsai art form, which was a reflection of these ideals.

The care of a bonsai plant wasn't just an aesthetic pursuit for Samurai. It was a symbol of the dedication they had to discipline and a skill. An experienced fighter needs the same level of determination in order to take a bonsai plant from its beginnings to an impressive work.

The Development of Bonsai Over Time

The transformation of bonsai as an image of wealth and power, to an enduring art form accessible to every walk of life been a significant influence on the development of the art discipline.

At first, bonsai was considered an exclusive privilege for the elite as a way to display their riches and status. As time went by the public began to take bonsai seriously as an expression of their own self as well as a means of connecting with the natural world.

The world is abuzz with Bonsai enthusiasts still adhere to the art's historical and intellectual roots even today. They experiment with different styles and varieties while respecting the tradition and pushing the boundaries of creativity.

In its long-lasting beauty and wisdom that is timeless Bonsai is truly spreading over centuries and across countries.

In this chapter, we will delve deeper into the philosophy and history of bonsai in this section considering its importance in culture and the spiritual lessons that it offers on a higher levels.

The foundations we will use for us with an understanding of the concrete elements of this art in terms of importance and context while we progress to the realm of bonsai.

Style: Informative and Educational

Illustration: A series of detailed drawings that show the process of growing a bonsai from seedlings to fully mature, stylish tree. It also includes subtitles for every step.

An example caption to describe images "Stage 1 - Sapling: A young tree is selected for Bonsai training, and its journey begins with careful pruning to establish the basic structure."

Chapter 12: Selecting The Best Bonsai

The second chapter will will begin an essential phase in studying bonsai, which is choosing the bonsai that is best. The entire experience of learning bonsai will be affected by this selection that will also have an effect on how the tree grows in appearance, how it looks and the way the relationship you have with it.

Then we'll explore the wide realm of Bonsai dimensions and styles as well as the finer points of picking the best kind of species.

An essential step of your bonsai experience is deciding on the right species dimension, design, and size. Learn in this chapter can help to make choices that align with your preferences and goals while we look into the various options.

2.1 Choosing the Ideal Species

Recognizing the Types of Species Choosing the most suitable Bonsai species is like selecting an ideal travel partner to help you in your artistic endeavors. The different species possess distinct characteristics as well as maintenance requirements as well as attractiveness. It is important to understand the details.

Climate and Environment The location that you will keep your bonsai as well as your local climate can play an important role when choosing the species. Some species thrive in climates that are tropical or subtropical Others thrive in more moderate environments.

Personal style Your vision of aesthetics for your bonsai aswell depending on your individual preferences are both equally important. Certain species are known because of their beautiful flowing branches. other species, thanks to their tall posture, exude energy. Find species that match your personal style is essential.

In terms of the amount of care required, different species need different levels of care. Some species require greater upkeep than others, a few have a strong and durable nature and are perfect for those who are new to the sport. We'll guide you to choose the best kind of animal for your level of ability.

2.2 Sizes and Styles of Bonsai

Finding harmony and a proper proportion the world of bonsai balanced and proportions are essential. Your bonsai's size will be in line with the look that you are planning to create. We'll provide more information about how you can make sure that your bonsai's visual elements are balanced in order to make it look good on all angles.

Learning to Get the Hang of getting the hang of Classics as the foundation for creative expression, a familiarity with the classic Bonsai style is vital. The following guide explains the nuances of each kind of style, such as the formal sophistication associated with the Formal Upright, the relaxed appeal that is the Informal Upright, and finally the stunning elegance that is Cascade. Cascade style.

The most important characteristics and strategies employed within these genres will be explained to the students.

Examining Special Styles (Continued) Beyond traditional types, bonsai promotes imagination and uniqueness. There are two distinct styles of bonsai. Literati and Windswept types of bonsai that defy convention and inspire creativity, will be explored in the following section. Be creative and look for inspiration in the unpredictably.

Giving a good first Impression Your bonsai's impact is able to make on the viewer will depend on the scale and appearance. It can express your style of art and personality regardless of whether you wish to portray strength, elegance or even the whimsy. We'll give you tips about how to choose the right size and style which effectively communicates your message.

The art of Bonsai styles The bonsai art form offers many different designs that each has a unique story to tell. The following section will focus on the styles that are most popular.

different styles of bonsai, including Cascade, Formal Upright, and Informal Upright. Learn more about the characteristics that define the different styles.

Style and size coordination bonsai's appearance and size are inextricably linked. To achieve the aesthetics required the various styles require the use of different diameters for trees. This article will guide you through the steps of choosing the best dimension to match your style preference.

Looking for Special Styles That Go Beyond conventional shapes, bonsai enthusiasts tend to experiment with more modern styles that push the boundaries of traditional art. To inspire you to further pursue your passion for art and interests, we'll help you discover certain of these unique techniques.

Bringing Peace Bonsai is centered around harmony. It is an art form by itself to ensure that the species is compatible in the most appropriate dimension and shape. If you want to ensure that your bonsai reflects the vision you have and is in tune with your soul, we'll provide practical tips for how to attain this balance.

This chapter will give readers with the details and knowledge necessary to pick the right bonsai plant with understanding and knowledge. The journey to the realm of bonsai begins with a meticulous selection of the perfect species, shape, and design, regardless of whether you're drawn by the elegance of a formal Upright design or intrigued by the enthralling possibilities of unconventional designs.

Style: Decision-Making and Guidance

Illustration: A visual representation of various species, sizes and bonsai types, as well as labels highlighting the distinctive characteristics of each.

A picture of the bonsai's species label is "Japanese maple (Acer palmatum) A favorite of traditional upright (Chokkan) style and is renowned for its bright fall leaves

Essential Equipment and Supplies

in Chapter 3, we dive deep into the most essential items and equipment every fan of bonsai must be armed with. These essential components form the base for bonsai maintenance and growth.

Wire and shears are trimmed, in three.

The bonsai sculptor's most trusted friend A Pruning Shear is one of the essential equipments to bonsai lovers is the pruning shears. Within the world of bonsai they're an equivalent to the sculptor's tool and painter's brush. It is possible to precisely trim the shape, create, and even perfect the bonsai tree using pruning shears. This article will examine a variety of styles of pruning shears such as concave and branch cutters as well as how to use them in particular ways for bonsai arts.

Practice Pruning in bonsai pruning is an art that requires a lot of skill. In order to achieve the proper shape and scale, unsuitable branches, leaves, as well as shoots need to be taken care of. The steps we'll show you are necessary.

The process of trimming your bonsai in a way to keep its beautiful appearance and its health. In order to achieve the results you want You'll be taught about techniques like directed pruning as well as the process of pruning buds.

Wiring: Designing the future of Your Bonsai Another essential part of bonsai art work is wiring. It allows you to design beautiful lines and angles which characterize Bonsai designs by gently steering branches to the correct position. In this article, we will discuss the different varieties of wire that are utilized in bonsai designs, like aluminum and copper in addition to the best ways to wire your tree securely and efficiently.

Pruning and wiring together A delicate balance you are growing your bonsai, wiring as well as pruning often are in tandem. For shaping your Bonsai tree and to ensure its health over time it is crucial to strike the right balance between these two ways. This article will provide suggestions about how to manage the trimming and wiring to create beautiful Bonsai compositions.

3.2 Soil and Container Choice

What is Foundation of Bonsai Health: Soil Composition It's crucial that you select the correct the soil that will support your bonsai. The tree's health is dependent on the soil's moisture supply and important nutrients for it's lifeblood.

Chapter 13: Potting And Repotting As Described

The most important aspects of the process of planting and the replanting of Bonsai trees are discussed in Chapter 4. These strategies are vital to keeping the longevity, growth and overall the health that your bonsai enjoys.

4.1 Transplantation Methods

Be aware of the necessity to transplant is an essential aspect of maintaining bonsai trees involves transplanting or repotting. When they age, bonsai trees grow out of their pots, use up the mineral content of the soil and begin to develop root disease. This article will examine the signs that your bonsai should be relocated to this area. It will help you recognize these signs and allow you to plan the tree's transplant precisely.

The practice of renewal The practice of renewal part of the root system in the process of transplantation is an intricate process referred to by the term "root pruning."

Though it might seem paradoxical the root trim is essential for bonsai's health.

In guiding you through ways of pruning your roots We can assist you to reach the right balance between removing extra roots and keeping the health of the tree.

Strategies for Potting: Making an All-New Home. To maintain the wellbeing of your bonsai tree, select the correct pot and employ appropriate potting methods. The discussion will cover the different potting techniques, like as the "slip-potting" approach and the "bare-root" approach. We'll give you tips on choosing the best pot shape and size to be a perfect match to the look of the bonsai you want to grow.

The Repotting Procedure How to Repot a Bonsai: Step By Step bonsai needs to be planted following a series of cautious steps. Repotting is an extensive process which includes everything from taking the plant from its current pot, to cutting its roots. This is a comprehensive guideline, step-by-step on how to repot to help you finish this essential task confidently.

4.2 The Right Time and Season

Consider the seasons when you are repotting your bonsai, timing is the most important factor. Each species has a specific development phase, repotting an individual tree in the wrong timing could result in stress. Here we will outline the ideal timings to plant your bonsai considering the species of it, its environment and stage of development.

Two main seasons for repotting that you'll study is early spring and late winter.

Repotting in the spring is a time to Renewal Repotting Bonsai typically occurs during spring. The growing season, during which trees are the most durable starts at this time.

In this article, we'll explore the advantages that repotting during spring time, including the development of roots and faster the ability to heal. We'll give you tips on how to prepare your bonsai prepared to enjoy a successful spring Repotting.

Repotting during the winter months or in the early spring is for Particular Cases Repotting during the later winter months or in the spring is sometimes necessary in certain species and under specific circumstances.

This article will give details about the circumstances that require repotting during this time and also the security precautions and requirements for maintenance of Bonsai trees transplanted in the winter of late or the beginning of spring.

Post-Repotting care: making the transition smoother The bonsai requires special care and attention following repotted to help make the transition smoothly. We'll provide advice for after-repotting maintenance such as fertilizing, watering and securing against harsh weather. These actions are vital to the health and rehabilitation of the bonsai.

Your Bonsai can continue to grow If you know the proper methods as well as the proper timing to plant and repotting. It will show a strong development and maintaining its appeal. These are the fundamentals of bonsai cultivation, which allows the owner to maintain your bonsai and watch the tree grow as time passes.

Style: Usable and Educational

Illustration: A neatly arranged image of the required Bonsai equipment and supplies along with callouts for the purpose of each piece.

Request in the case of Bonsai Wire, for instance "Bonsai wire offers flexibility and control during styling and is used to shape branches and maintain desired positions."

Bonsai Maintenance and Care

Chapter five's focus is the best way to care and maintain the bonsai tree. This is essential to ensure the bonsai's longevity energy, vitality, as well as aesthetic attractiveness.

Feeding and Watering

Water's Function in Bonsai Health

One of the most essential aspects of Bonsai treatment is to keep it watered. In this section we'll look at how crucial water is in keep your bonsai in good health. We'll discuss the specific requirements to water bonsai plants which include factors such as soil composition temperatures, species specific requirements.

The Art of Watering, Perfected

The proper watering of a bonsai is a skill. It involves determining the proper amount of water to your plant while also making sure to avoid overwatering that can result in root decay. The following article will provide helpful tips regarding how to keep your bonsai hydrated, with methods such as using the "soak and dry" method as well as the use of humidity tray.

How to Feed Your Bonsai: Fertilization Methods

A further important aspect essential to Bonsai maintaining is fertilisation. Since bonsai plants depend on the soil's minerals, the nutrients gradually depleted over the course of time. To maintain your bonsai's strength and health, we'll walk through the different types of fertilizer that are available, both organic and chemical options, as well as provide advice regarding when and how you should fertilize your bonsai.

Pruning and wiring

The importance of trimming for health as well as appearance

Pruning in bonsai can be both an art form and it is a science. For you to form your tree's shape and encourage the growth of your tree, it is necessary to do careful pruning of branches and leaves. The following article will discuss pruning strategies like pruning for structural purposes to enhance the shape of your tree as well as maintenance pruning that will enhance its appearance. Also, you'll be able to identify the tools and time required to ensure efficient pruning.

The Sculptor's Tool connects to the electrical grid.

Bonsai trees can be styled and designed using a technique known as wiring. Wiring lets you define the style of your bonsai tree by twisting branches, and creating beautiful curvatures. In this article, we'll provide more information regarding wiring techniques including how to put in and remove wires to ensure that the branches of your tree are placed in a harmonious manner within the design you prefer.

Prevention of Disease is Important

Chapter 14: Tuning In To The Nature's Rhythms Using Seasonal Care

The preparation for the winter season is called Dormancy. The Bonsai calendar Winter is the most important time of year. A lot of Bonsai species enter hibernation during the time of hibernation. The following article will cover the special requirements for winter maintenance like securing the plants against freezing temperatures, providing adequate light and changing irrigation and fertilizer schedules.

The bonsai can come out of its dormancy with good health provided you provide it with the appropriate winter-time care for it.

The arrival of spring Promotion of New Growth bonsai's will come out of winter hibernation to begin a new development cycle when spring gets closer. We'll assist you in making changes, such as the need to repot your tree if needed and altering the frequency of maintenance to accommodate the increased needs for nutrition and water.

With the growth that is fresh on the bonsai's branches spring is also an ideal season to design and install wiring.

Summer Be Watchful: Juggling Growth and Pruning Most bonsai species are most active during the summer months.

You'll be required to trim and wiring to ensure the strength of the tree.

This article will discuss methods of maintaining your bonsai's form and healthy throughout the summer, while ensuring it receives enough light and water.

Autumn Transition: Winter Preparation

The patterns of growth and color of leaves will change when fall is near. Let's look at ways to manage this transition and make your bonsai ready for next winter. It covers methods of gradually cutting down on watering needs and securing your bonsai from changes in temperature.

It is now clear that you have a good knowledge of how you can take care and manage your bonsai starting with the basic tasks such as watering and feeding, to the seasons-specific factors that ensure your tree's wellbeing all through the year. If you implement these strategies in practice, you'll not just take good care of your bonsai's aesthetics and health as well as strengthen your connection to this contemporary and timeless method of gardening.Style In-depth care guide

Illustration: A style of calendar showing the requirements for seasonal maintenance for Bonsai and includes symbols to indicate each month's watering and trimming, feeding and prevention of disease.

As an example for July's month, its image of a watering container as well as a set of scissors reads "Water deeply, and perform maintenance pruning to maintain shape."

Fundamental Bonsai Styles

Learning how to understand the Bonsai Styles' Foundations The shape and character of your tree can be visualized through the bonsai style. The most important Bonsai styles that provide the base for creativity are discussed in this chapter. The styles include:

Traditional Upright (Chokkan) The design is a symbol of strength and stability because of its straight, upright trunk.

An informal Upright (Moyogi) The Moyogi style has a relaxed and organic look due to its gently curled trunk.

Cascade (Kengai) The model resembles a tree growing out of the cliff and its branches cascading down.

Slanting (Shakan) A slanting style features a trunk with a diagonal which symbolizes adaptability and sturdiness.

Windswept (Fukinagashi) The name suggests, this pattern is reminiscent of the look and feel of a tree's branch that are being swept in only one direction by the force of winds.

Literati (Bunjin) Literati (Bunjin) style is a symbol of artistic expression and innovative thinking. It is distinguished by its asymmetrical, deformed look.

Utilizing the group planting (Ikadabuki) method A small or a landscape is created by putting together a number of trees.

Picking the Bonsai Style The most appropriate style to choose for your bonsai should reflect both your artistic vision and the intrinsic attributes that the tree has. The following article will provide suggestions about how you can combine the traits of your tree to create a style of bonsai that is appealing to your taste. We'll show you how to follow the guidelines of each design while also enhancing the beauty of your tree.

Complex Shaping Methods

Develop Artistry by Using Modern Methods Innovative shaping methods allow users to explore the limits of expression through art regardless of whether understanding the basic techniques is essential. In this article, we'll discuss some strategies that can help the bonsai to reach the status of a masterpiece:

Jin Shari Shari The purpose of these procedures is to produce deadwood traits such as the jin (dead branches) and the shari (deadwood on the tree's trunk) This process involves cutting off of wood and bark. These give trees character as well as age.

Bunjin (Literati) Style The sophisticated style encourages differentiating from the norm, and a creative approach. This article will explore the Bunjin design principles that permit unique and avant-garde Bonsai forms.

By layering, you can make it appear as if you have several species of trees that grow from one. The layers give your bonsai arrangement an extra dimension and depth.

As with group planting it also has a variety of trees. The placement of the trees is carried out so that it appears more real. The clumps look like trees being cultivated in thick forests.

The Style of the Bonsai Changes with the season. When your bonsai gets bigger and season changes, styling and shaping is a continuous process that also change. Learn how to change your methods of styling all year long to ensure that your bonsai is an exciting and captivating piece of art.

In this chapter, you will be able to study the most essential styles as well as sophisticated technique for shaping, you'll be able to learn more about the art of bonsai within this chapter. It will enable you to better appreciate it. The place where the individual style of your aesthetic as well as your bonsai's style connect is with the way you style and shaping of your bonsai, no matter if you wish stick to the established style or break the boundaries of creativity.

Design: Creative Presentation

Illustration: An image of a well-set up bonsai display, with decorative objects, and just the right quantity of light to provide the perfect atmosphere.

Example: "A Bonsai display featuring a Formal Upright style tree, complemented by a kusamono accent plant, creating a harmonious composition."

A Display of Your Bonsai

Displaying your bonsai trees in public view is explained in Chapter 7 in which the author also stresses the significance of displaying and participating at bonsai exhibitions and shows.

Presentation and Display

The bonsai exhibit is a piece of art by its own. The effect of the tree's appearance increases and guests are capable of fully absorbing the beauty of your tree and its craftsmanship due to the display. The following article will discuss the key elements of bonsai displays in this segment:

The selection of the right display stand or table is essential when it comes to bonsai. We'll discuss a variety of options and assist you in selecting one that is appropriate to the dimensions and design of your bonsai.

Companions and accent plants Accent plants, often known as kusamono, or shitakusa give your bonsai displays the context and dimension it needs. This article will discuss the most effective methods to select and set up these companion plants.

Show Seasonality: The plant's beauty can change according to the seasons, thanks to bonsai displays and accentuates the fact that this. Learn how you can alter your display in line with your season as well as the emotion that you'd like to convey.

Bonsai Shows & Exhibitions

The Exhilaration of Bonsai Exhibits By taking part in bonsai exhibitions or events, you will be able to connect with fellow fans, and show off your artistic ability. This article will examine a variety of bonsai exhibitions like:

Your bonsai needs to be prepared carefully if you plan to exhibit it. For the sake of ensuring that your tree is in good condition to show We'll provide an outline of the steps.

Show Etiquette: There are a variety of rules and guidelines that go along when you exhibit your bonsai. Learn how to conduct polite and professional interaction with judges, other participants, as well as members of the public.

Networking and Learning: at bonsai events the opportunity to network and learn from the other attendees is just as important as showing off your tree. The discussion will focus on the benefits attending exhibits as well as performances, from getting invigorated to connecting with peers and mentors.

How to Prepare to Attend your First Exhibition If you're getting ready to host your first Bonsai show, we'll guide you through the process as well as offer tips to ensure your debut be successful.

The confidence you gain will be the skills to bring your bonsai to the wider community of bonsai enthusiasts. your bonsai.

Learn how to present your bonsai's work to its greatest potential in this section. How to present your bonsai and exhibiting etiquette will assist in expressing your enthusiasm and creativity to other members of the bonsai world, regardless of whether you want to create stunning displays to decorate your home or to participate in shows or shows.

Design: Creative Presentation

Illustration: A photo of a well-set up bonsai display, with accessories, plants and just the right quantity of light to provide an inviting atmosphere.

The Explanation: "A Bonsai display featuring a Formal Upright style tree, complemented by a kusamono accent plant, creating a harmonious composition."

Chapter 15: A Source Of Serenity

Chapter 8 focuses on the huge impacts that bonsai could affect your health as well as the ways it will help you live in peace and understanding.

Mindfulness and stress management

The Bonsai's Healing Power

An unique method of relaxation and mindfulness that bonsai grows provide is. This article will explore the benefits of taking care for your bonsai within this segment, including:

Meditation: Patience, patience and focus on the details are crucial to cultivate bonsai. The discussion will focus on integrating meditation exercises within your bonsai practice for a more peaceful and centered experience.

Reduce Stress: Working with your bonsai, irrespective of whether you are trimming it wires, tying it up, or simply taking a look at it grow, can provide a relaxing escape from stress and pressures of daily life. Learn the ways Bonsai can help you relax and rejuvenate.

Nature Connection Bonsai provides a profound relationship to nature. This link can make to feel grounded, and bring peace and harmony to our chaotic modern day.

Fostering Calmness at Home

Including Bonsai in Everyday Life

It is possible to regularly encourage tranquility and peace by establishing the Bonsai space in your home. The following tips will help you in adding bonsai to your house such as:

Selecting Indoor Bonsai Indoor Bonsai could bring the beauty of nature into your home the home if you are in a place with extreme winters and limited outdoors space. In this article, we will talk about the indoor species of bonsai that can be used for maintenance and appropriate techniques.

The use of bonsai for decor could be used to create a beautiful décor, improving the look of your home and providing the ambience of a calm. It will be clear how to pick the best bonsai to suit different areas and styles of architecture.

If you're blessed with outdoor space, creating an area for a bonsai tree could change your life. This guide will help you design and manage a peaceful outdoors space stocked of bonsai trees.

Educating Others About Bonsai

The advantages of sharing the bonsai love to other people is innumerable. The discussion will focus on ways to introduce bonsai to family and friends by offering classes or the creation of a local bonsai community.

Accept Bonsai Beauty In Your Life to conclusion.

It is a process that involves acceptance of bonsai as an element of calm and peace in your daily life, in the final section. Bonsai is not just an exercise for gardeners. it's a method to grow oneself and to find peace within.

There is a chance to find consolation in the form of inspiration, motivation, and eternal beauty in this art by integrating the bonsai knowledge into the everyday.

Stress reduction and improved mindfulness

The care and maintenance of bonsai trees are natural nature observant practices. When you observe your bonsai trees grow, take precise cutting cuts or connect its branches to the ground and forces you to stay fully present and present in the moment.

The ability to put off temporarily your worries and stresses that are part of everyday life due to the consciousness that creates.

Engaging with your bonsai can greatly reduce stress According to studies. There are times when you notice that the stress and anxiety go away when you are immersed in the relaxing routine of maintaining your bonsai. It takes a lot of patience to care for living things can provide a tranquil escape away from the stress of a hectic and fast-paced world.

Connecting to Nature Connecting with Nature: Even in urban environments, bonsai serves as a way to connect with the natural world. Within an era of modernity driven by technology, and often detached, this connection with the natural world could bring back your senses and make you be in a place of calm.

There is a sense of calm and a sense of perspective when you keep in mind the progress and seasonal variations of your bonsai plant and is a permanent reference to the cycles of nature.

Fostering Calmness at Home

If you live in extreme weather areas or are unable to access to outdoor space Indoor bonsai can allow nature to be brought indoors. The capacity of a species to adjust to conditions in indoor environments is an important factor to consider in deciding whether to use them as indoor bonsai.

They add a touch from nature to your house and serve as a point to reflect and relax.

Bonsai for Decor Apart from being art pieces they can also be used as decor for your home. Selecting bonsai that match your style can improve the overall harmony. The bonsai can be seen or even heard.

Beautifully attractive and emotionally soothing creating your living areas to tranquil havens.

Building outside Sanctuaries: if you're lucky enough to enjoy outside space, constructing an outdoor bonsai planter could transform your life. It is possible to escape into nature and not leave your home in the form of an outdoor oasis that is populated with Bonsai trees.

They provide places for meditation and relaxation that allow you to connect with nature, and admire bonsai's beauty.

Introducing Others to Bonsai

Introduce your family members to Bonsai could be an enjoyable adventure. Organising local Bonsai groups, or facilitating workshops can help foster connections and sharing of knowledge. This allows you to feel the peace and calm which Bonsai can bring to others and creates a sense of belonging and growth.

Mindfulness and stress reduction

The care and care of bonsai trees are naturally attentive nature practices. When you observe your bonsai tree expand, take precise pruning cuts or gently connect its branches to the ground and forces you to remain completely and present in the moment. The tree will temporarily let go any worries or stressors that arise in everyday life because of the sense of awareness it creates.

Engaging with your bonsai can help reduce stress levels According to studies. It is common to see the stress and anxiety go away while you take a deep dive into the relaxing patterns of maintenance for your bonsai. It takes a lot of patience to care for an animal is a relaxing escape from the stresses of a hectic and fast-paced world.

Connecting to Nature In urban areas, bonsai can act as a means of connecting to the natural world. While living in our modern society, immersed in technology, and often absent, this connection with nature may revive you and make it easier to be at peace. It is possible to find peace and perspectives through keeping track of the changes and growth of your bonsai and acts as a permanent reminding you of the cycle of life.

Fostering Calmness at Home

The indoor bonsai can bring the outdoors indoors to those living in extreme weather areas or are unable to access to an outdoor area. The species for indoor bonsai have been specifically selected to be able to be adapted to indoor conditions. They create a point to relax and contemplate and also bring an element of nature to your living space.

Bonsai as decor: Bonsai may be used as decor for your house and are also pieces of art that live. Selecting bonsai that are in line with the style of your home can help to create peace and harmony. They can transform your living spaces tranquil refuges, as they are both visually attractive and emotionally soothing.

Building outsidesanctuaries If you're lucky enough to enjoy an outdoor space, creating the bonsai gardens could transform your life. It is possible to escape into nature in your own home without ever leaving it in the form of an outdoor sanctuary by using Bonsai trees.

The gardens provide spaces that allow you to contemplate and relax that allow you to connect with the natural world and enjoy bonsai's glory.

Accept Bonsai Beauty throughout your life in conclusion.

We encourage you to contemplate the positive impact that bonsai can have on your life during the final section. Alongside having a hobby in the field of horticulture, it is also an effective method to gain mental stability, personal development and strengthen your connection to nature. It is possible to find the comfort of inspiration, peace and peace of mind with this timeless and beautiful approach by accepting the beautiful of bonsai, and incorporating the wisdom of bonsai into your daily day life. Alongside the care for trees, bonsai also involves improving personal wellness and finding calm in the chaos of every day living.

In conclusion this way, the practice of bonsai offers an opportunity to be awed by the peacefulness, peace and peace that this ancient art forms offers, rather than the mere growth of tiny trees.

Bonsai is not just about gardening. It's an opportunity to build an even closer connection with the natural world and to help find peace amid a chaotic life.

In launching this bonsai experience by registering, you're beginning on an adventure that can lead you to:

Through a call to you to be mindful as taking care of the tree, bonsai encourages meditation. Bonsai is a place that is focused and present in an environment that is rife with distractions.

The ability to care for an animal and watching its growth can provide a relaxing escape from stress and pressures that come with everyday life. A bonsai-garden can provide an idyllic setting to rest and relaxation.

Connectivity with Nature: Bonsai re-establishes your relationship to nature, by the beauty and rhythms of nature to urban spaces. It creates an ambiance of harmony with the environment.

The Living Spaces that are Calm: Bonsai can help you make living spaces tranquil and gorgeous inside your house. A work of art, and an instrument of happiness the bonsai.

Sharing and community Sharing and Community rewarding to pass on your enthusiasm of bonsai among other people. Organising workshops, joining local associations, and getting to know fellow bonsai lovers create an atmosphere of community and create a sense of mutual growth.

Then you embark on an adventure of inner peace discovery, self-discovery, as well as the development of your own self as you accept the beauty of bonsai throughout your daily life. It serves as a constant reminder of the beauty within the outside and within oneself, bonsai becomes the source of relaxation as well as inspiration and lasting calm.

Keep in mind that bonsai can be used to dramatically improve the quality of life regardless of your previous experience level with the captivating art of. It's a chance to slow down and enjoy the splendor of your present and build a deeper connection with nature. Let the tranquility that comes from bonsai's beauty in your daily routine, and let it transform the direction of your journey.

The cultivation of Bonsai trees is a tradition that has been in existence for more than 1,000 years. Most people do not realize about the precise meaning for Bonsai, "Potted Plant," is simply a potted plant. But, there are many varieties of trees, plants and even trees could be maintained and cultivated by bonsai.

The cultivation of bonsai trees is often associated with Japan however, the practice began in China and the area which is where they later became closely associated with Buddhism. Zen Buddhist faith. Alongside their more traditional use, bonsai tree are now being used for decorative as well as recreational reasons.

When maintaining bonsai plants, the person who cultivates them is able to be a creative and thoughtful contribution to the growth of an unadulterated symbol beauty.

7 WAYS TO GROW A BONSAI BEAUTY SERENITY Part 1

Selecting Your Ideal Bonsai Tree

Pick a tree type suitable to the climate. There are a variety of bonsai plants. There are even certain tropical plants, and some woody perennials could become bonsai trees however not every species can thrive in the specific environment you reside in.

The climate where the tree is to be grown is a factor to consider in deciding on a specific kind of tree. Certain species are damaged by frigid temperatures, while some require the temperature to be lower than freezing to be dormant to get ready to spring.

Check that the plant you've selected is suitable for the location you're in prior to planting the Bonsai tree, specifically in the event that you plan to plant an outdoor tree.

If you're unsure, contact employees at your local gardening supply store for advice.

The juniper is one of the Bonsai species that's great for those who are new to the art. They are abrasive that thrive in the northern hemisphere as well as in the southern hemisphere's less humid regions.

Furthermore, they can be grown easily because they're evergreen and do not lose leaves. They can also be tolerant of pruning as well as various other "training" techniques. Their growth, however, can be a little slow.

Conifers which are commonly planted as bonsai plants include various species of pine, spruce and cedar. A different option is to cultivate trees that are deciduous (leafy) trees. Japanese maples and magnolias as well as and oaks are especially beautiful.

The last but not least is that tropical plants that aren't woody such as jade or "snowrose" are suitable for indoor conditions in cooler or warm regions.

Part 2

Choose whether you would prefer an indoor or outside tree.

Bonsai plants that are indoor and outdoors could have very different needs. It is recommended to select trees that have less moisture and light requirements since indoor areas are usually more humid and receive less sunlight than outdoor ones.

The most sought-after bonsai tree varieties are listed below. classified according to whether or not they can be used in outdoors or indoor settings:

Inside: Kingsville Boxwood, Hawaiian Umbrella, Serissa, Gardenia, and Camellia.

Outdoors: Elm, Ginkgo, Larch, Maple, Birch, Cypress, Juniper, and Cedar.

Be aware that without winter dormancy the temperate species will eventually end up dying. They are not able to be grown over long durations inside.

Part 3

Select the bonsai's size. The sizes of bonsai plants can vary widely. Based on the type of tree the full-grown tree can vary between six inches (15.2 cm) up to 3 feet (0.9 meters). The size of the trees can be reduced in the event that you want to plant your bonsai using a seedling, or a branch from another. Be sure to have all necessary resources prior to purchasing because bigger bonsai trees need greater amounts of water, soil and sun.

Take into consideration these factors when deciding on the bonsai size for your tree:

The container you'll use's dimensions

The space you are in at your home or in the office

Sunshine is a welcome sight on your property or in the area of work

The amount of care you can give the tree (bigger trees need more time to trim)

Part 4

While selecting a plant visualize the finished item in your mind.

It is possible to visit a nursery or bonsai retailer to select the tree which will develop into a bonsai plant when you've determined the kind and size of bonsai you'd like to have.

For a way to determine if a plants is healthful, look for vibrant, healthy green needles or leaves (but be aware that deciduous trees could be different colored leaves during the autumn).

After you've narrowed down your choices to those that are healthy and most beautiful plants, think about what each one will look like after trimming.

The process of gently shaping and trimming a bonsai plant until it's exactly the way you would like it to be could require a long time, but that is an integral part in cultivating a bonsai tree.

Select a tree whose shape is suited to your plan of shaping and pruning. (Many bonsai were specifically designed to mimic the proportions of form, shape, and leaves of mature trees.)

You will control the growth of your bonsai tree at nearly all stages of its development, when you decide to plant your Bonsai plant from seeds.

For a bonsai plant to mature from a seed to fully mature trees, it can take between five to 5 years (depending on the species that tree). This is why it is recommended to purchase mature plants if you're looking to instantly trim or form your tree.

There is also the option of establishing your bonsai plant by cutting it.

Cuttings are a type of branch that is harvested from the trees they are placed in new soil to form a new species that has a genetic similarity to the one that was originally.

www.ingramcontent.com/pod-product-compliance
Lightning Source LLC
Chambersburg PA
CBHW071440080526
44587CB00014B/1921